MW01627596

THE SOCIAL CODES OF TECH WORKERS

LABOR AND TECHNOLOGY

Winifred Poster, series editor

Madison Van Oort, *Worn Out: How Retailers Surveil and Exploit Workers in the Digital Age and How Workers Are Fighting Back*

Sofya Aptekar, *The Green Card Soldier: Between Model Immigrant and Security Threat*

Margaret Jack, *Media Ruins: Cambodian Postwar Media Reconstruction and the Geopolitics of Technology*

Coleen Carrigan, *Cracking the Bro Code*

Daniel Goodwin and Edward Schwarzschild, *Job/Security: A Composite Portrait of the Expanding American Security Industry*

Robert Dorschel, *The Social Codes of Tech Workers: Class Identity in Digital Capitalism*

THE SOCIAL CODES OF TECH WORKERS

Class Identity in Digital Capitalism

ROBERT DORSCHEL

The MIT Press
Cambridge, Massachusetts
London, England

The MIT Press
Massachusetts Institute of Technology
77 Massachusetts Avenue
Cambridge, MA 02139
mitpress.mit.edu

© 2025 Massachusetts Institute of Technology

This work is subject to a Creative Commons CC-BY-NC-ND license.

This license applies only to the work in full and not to any components included with permission. Subject to such license, all rights are reserved. No part of this book may be used to train artificial intelligence systems without permission in writing from the MIT Press.

This book was set in Adobe Garamond and Berthold Akzidenz Grotesk by Westchester Publishing Services. Printed and bound in the United States of America.

Library of Congress Cataloging-in-Publication Data

Names: Dorschel, Robert author.
Title: The social codes of tech workers : class identity in digital capitalism / Robert Dorschel.
Description: Cambridge, Massachusetts : MIT Press, [2025] | Series: Labor and technology | Includes bibliographical references and index.
Identifiers: LCCN 2025017042 (print) | LCCN 2025017043 (ebook) | ISBN 9780262553537 paperback | ISBN 9780262384940 pdf | ISBN 9780262384957 epub
Subjects: LCSH: High technology industries—Employees—Social conditions—United States. | High technology industries—Employees—Social conditions—Germany. | High technology industries—Employees—Economic conditions—United States. | High technology industries—Employees—Economic conditions—Germany.
Classification: LCC HD8039.H54 D67 2025 (print) | LCC HD8039.H54 (ebook)
LC record available at https://lccn.loc.gov/2025017042
LC ebook record available at https://lccn.loc.gov/2025017043

EU Authorised Representative: Easy Access System Europe, Mustamäe tee 50, 10621 Tallinn, Estonia | Email: gpsr.requests@easproject.com

For my parents Martin and Susanne

Contents

Prologue

Some people regard Austin as the new Silicon Valley, Rachel tells me at the start of our interview. Rachel is a twenty-nine-year-old data scientist based in Austin, Texas, who began her professional career as a risk analyst at a big bank before transitioning to the start-up economy. Even though we have no connections in common, Rachel immediately and warmly agreed to be interviewed for my study when I contacted her through LinkedIn. I'm reminded of this friendliness when she talks about the work culture at her office, painting a picture far removed from the competitive, individualistic tech world we might imagine:

> We definitely have good, like, teamwork and solidarity on our team. You know, people are very friendly and very willing to help each other. We, like, review each other's code, and people will post, like, "Hey, I'm trying to do this. Does anyone have any ideas?" and, you know, people will chime in. So I'd say it's very, you know, collaborative.

Rachel,[1] who identifies as a white woman, holds a master's degree in statistics and managed to learn most of the necessary computer skills for data science by herself online. Now a lead data scientist responsible for the analysis of large datasets and the development of algorithms at an insurance tech start-up, she earns $165,000 per annum. Rachel is part of an occupational segment that has benefited from the intensified digitalization of contemporary societies. This segment of so-called tech workers can be understood as the professionals in charge of programming, designing, and managing the digital technologies and services that permeate social space. These professionals bring into being the platforms, algorithms, and artificial intelligence (AI) applications that have become increasingly intertwined with our everyday lives. Their feats of technological engineering shape what we buy, where we go, and whom we trust.

In the modern tech economy, digital expertise is currency. Tech firms are hungry for professionals who can help to them to develop and further commercialize information and communication technologies: Data scientists, software engineers, product managers, AI developers, user experience (UX) designers, and other tech work professions are in high demand. Despite recent layoffs, this professional segment has grown significantly since 2018, when the Computer Technology Industry Association (CompTIA) began tracking the field. While AI may change things, current numbers remain stable, with CompTIA (2025) reporting that 2.36 million technology professionals are employed by tech companies in the US—roughly twice the number of high school teachers (US BLS 2023). Due to the high demand for digital expertise, it is not uncommon for newly hired employees to command six-figure starting salaries (especially in the US) as well as an array of perks, from wellness programs to stock options. In general, tech workers occupy an economic position somewhere in between the slim elite class of tech entrepreneurs and the vast lower class of highly precarious digital workers such as gig and crowd workers. Their economic position can be understood as a fusion of bourgeois and proletarian characteristics, resulting in a situation where they face conflicting economic forces (Amrute 2016; Irani 2019; Tarnoff 2020; Wright 1976). This constellation raises vital questions about how tech workers negotiate their contradictory situation. What drives those who build and maintain the platforms yet hold no meaningful ownership of them? What kind of identities does the digital middle class cultivate?

Conventional wisdom might suggest that this kind of well-off professional group would embrace the economic status quo. Consider bankers, for instance, who have reaped significant benefits amid financialization since the 1980s. We rarely think of them as social critics—and for good reason, as sociological studies show a pronounced market-oriented habitus among financial professionals (Neckel et al. 2018; Neely 2022). With tech workers, though, a different story emerges. Rachel and most of the other tech workers I interviewed between 2020 and 2022 bear little resemblance to the wolves of Wall Street in their self-presentations. Rachel doesn't come across as someone who has achieved the American Dream by subordinating her life to market logic. Her heart and mind are not hardwired to the beat of the market. Instead, she is quite eager to distance herself from capitalist thinking throughout our interview:

> I'm definitely for critiquing the tech industry and corporations, capitalist entities in general. Just because I work for one and that's my job, like, I don't think that my company is exempt from any of that. And I think people who work for companies need to hold their own companies accountable, like people have done at Google with protesting and things like that. So, you know, I like my company, they've been good to me so far. But I'm not, sort of, loyal no matter what they do or anything like that. And, you know, I don't live to be a data scientist. So yeah, that's my experience. I think a lot of a lot of people are sort of like me.

Rachel holds a critical stance toward both her employer and her professional role. She maintains a deliberate distance, rejecting the notion that work should define who she is. We would be misunderstanding Rachel if we categorized her as a straightforward "entrepreneurial self" or "*Homo oeconomicus*," though these concepts have been the go-to terms for many social scientists seeking to label the subjectivity of knowledge workers (Bröckling 2015; Brown 2015; Reckwitz 2020; Skeggs 2004, 74). My research indicates that these generalizing categorizations of an (upper) middle class fundamentally bound up with neoliberalism are misleading. Like many of my other interviewees, Rachel seeks to cultivate a return to social critique, and she perceives herself as part of a collective with similar worldviews. She is not a continuously self-optimizing agent without a moral backdrop but is instead deeply skeptical of contemporary capitalism. At the same time, Rachel is not a revolutionary. She does not fundamentally oppose market principles:

> I really like my company, because, I mean, nothing's ever like a meritocracy, really. But I do feel like I've been rewarded for hard work and good ideas, and, like, I've been rewarded for, you know, the things that I've done, which makes me want to stay because, like, it proves if I do good work in the future, I'll also get rewarded for that, like, the incentives are sort of aligned.

Despite her social critique of the tech industry and capitalism in general, Rachel is attracted by certain facets of the market. She values merit and appreciates being rewarded for hard work; after all, she tells me, she invested a lot of time and energy into getting to where she is today. There's a quiet satisfaction in her voice as she describes how she recently took out a loan to buy a house: "Austin is expensive because it is a big tech market now, so it was definitely sort of, like: Buy a house before it becomes too expensive to buy a house. Because our housing market is like that, for sure."

Rachel is thus best described as a *hybrid.* She should be understood as someone who seeks to separate herself from the operations of capitalism without occupying a genuinely anti-capitalist identity. She strikes me as a person who wants to be both *middle-class wealthy* and *morally worthy.* On the one hand, Rachel's life conduct is governed by social codes associated with achieving and securing a comfortable standard of living. She cherishes hard work, has a mortgage to pay, and appears rather risk averse. On the other hand, Rachel remains committed to values that center on social justice and the welfare of others. Rachel aspires to be middle class without losing her moral compass—without becoming absorbed by the market, without compromising her principles. Though the tech workers I interviewed varied in many ways, this particular form of hybridity—and accompanying inner conflict—emerged as the striking commonality among them.

For example, across the Atlantic, in another tech hub, I spoke with David, a white Irish tech worker based in Berlin. David shapes the architecture of our digital lives as a UX designer; he is responsible for the usability and appeal of digital products, including their virtual interfaces. I made contact with David through a coworking space in Berlin-Mitte, where an old brick building has been transformed into a vibrant space where young, stylish professionals gather around the glowing screens of their high-end digital devices. When we did the interview, David was at home in his three-bedroom apartment in Neukölln-Schillerkiez, a trendy neighborhood comparable to Williamsburg, New York. David is thirty-seven and earns €80,000 a year. Wearing a plain white t-shirt, he comes across as polished but ordinary. When I ask him if he could imagine working for a big tech company on the US West Coast, he does not dismiss the proposition but is nevertheless eager to tell me:

> If money was the only thing that I was interested in, I would have become a dentist a long time ago. I mean, that's not my driving factor. You know, I mean, I want to live my life, and I want to do it in accordance with my values and to, you know, to also feel like I am contributing something.

David also defies simple categorization as an entrepreneurial self. While less explicitly critical of capitalism than Rachel, he is far from an opportunistic "surfer" willing to ride any wave that will take him to the top. Like Rachel, David's conduct of life reflects moral considerations; he is guided by other-oriented values manifest in his concern for other people's lives. Following Weber (1994), we may say that when he talks about his "Beruf" (occupation), he embodies a "Berufung" (calling). Notably, this calling is not channeled into the pure technological "solutionism" typical among the news-dominating class of tech entrepreneurs (Morozov 2013; Nachtwey and Seidl 2020). David doesn't view technology as an all-purpose hammer that can fix any social problem with a simple strike. Instead, he talks about "ethical design" and shows an honest interest in reflecting on the potential harms of digital technologies. His concerns mirror public debates about the negative effects of digital technologies. Notably, the ethic of care and responsibility that David cultivates is reinforced by a sense of confidence and collectivity. David considers himself part of the tech worker community and attributes power to this grouping. Consider this exchange:

> **David:** This emerging industry has changed the power dynamic. Power lies more in the hands of people like me than it did before. Of course, it's not fully in my hands, but it lies more with technicians, the engineers and designers than with the managers for the first time.
>
> **Robert:** How so exactly?
>
> **David:** Because they don't have a fucking clue what we're doing. If we left they couldn't, they wouldn't have, you know, they don't code, they don't design, they don't understand what we're doing. They're being brought in by Siemens or Volkswagen or whatever to be a stable head of the ship. But they could be thrown overboard, and nothing would happen. They are in a very precarious situation.

David does not see the world as a flat network. He thinks of society as conflictive and is reflective about power positions, though he feels a great of deal of confidence and security regarding his own position in the labor market. This confidence acts as a stabilizing element for his other-oriented values, which, my research suggests, are an integral part of the class formation of tech workers. Further research is necessary to evaluate to what extent this confidence, and other associated social codes, are retained in light of recent layoffs, advances in AI, and political shifts. What is clear is that in the early 2020s, the social codes of tech workers frequently translate into "boundary-makings" (Lamont 1992) vis-à-vis the entrepreneurial and managerial class. That is, they cultivate a distinctive identity based on ideals of reflection, collectivity, and solidarity.

Of course, an exhaustive class analysis would have to relate an analysis of self-understandings to material processes (Wright 2004), while this book limits itself to analyzing the hearts and minds of tech workers. Though my empirical analysis does not include ethnographic observation, the interviews nonetheless allow me to take reports of practices into account. For instance, even though most of my interviewees told me they could imagine joining a union, only two of the fifty-two reported currently belonging to one. Of course, union membership is not the only form of collective action, and subjective reports of practices only offer glimpses into the complex relationship between what actors say and what they do. Furthermore, a few of my interviewees did report engaging in practices of workplace solidarity, such as informally establishing salary transparency among colleagues to strengthen their bargaining position with their employers. Nonetheless, it was striking how rarely even strongly articulated moral values and social critiques matched with the concrete practices reported (a tension that is not unique to tech workers). To understand why, we must look deep into the hearts and minds of this contradictory class. The accounts in this book show that tech workers often remain tied to individualist ways of thinking and seeing the world, despite their widespread yearning for collectivity. While they generally want to alter the way the tech industry works, they typically ask what "*I*" rather than "*we*" can do to change it. Moreover, their dissatisfaction with capitalism exists within a wider subjective framework of disillusionment in terms of imagining other modes of economic and social organization. These subjective orientations partially explain why there were rather few protests against the massive layoffs at "Big Tech" firms in 2022 and 2023, or against the recent collaborations between tech industry leaders and Donald Trump. While researchers documented a major growth in tech worker activism between 2017 and 2019 (Tan et al. 2023), signaling important shifts within the tech industry, their data indicates that this momentum has stalled, with a substantial decrease in reported tech worker collective actions from 2021 to 2022.

The Social Codes of Tech Workers explores the contradictory class identity of these professionals and how they connect to urgent questions of power within our increasingly tech-mediated present. This book invites us to consider the intricate dynamics of subjectivity and structures of domination present within the world of tech by bringing into focus this influential yet constrained and only partially realized class of workers, whose mindsets and choices influence not only their economic field but our everyday lifeworlds. Along these lines, I also consider how the tech industry remains a work habitat with significant barriers for marginalized social groups, despite valorizations of diversity (see also Amrute 2016; Hicks 2017; Irani 2019). The experiences of Dalia, one of my interviewees from Berlin, illustrate this point. Born in the United Arab Emirates, Dalia identifies as a Black woman and reports facing multiple hurdles based on these characteristics, reflecting that "in tech, you will see more female UX designers than male UX designers. But the more famous UX designers are typically male." Furthermore, she tells me that designers are generally attributed a lower professional status than others in the world of tech:

> So I used to manage a blockchain meetup. [. . .] I wanted to bring, you know, thought leaders together. And not only focus on technology but also talk about policies and design. And I always, literally, every single time I hosted an event, I got the question "Are you a developer?" or "Who authorized you to take over this meeting?" There was always this really bad behavior from blockchain developers [. . .] Engineers are always treated like their gods. And it sucks that it's only the male engineers that are treated like gods not even the female engineers. And especially as a Black woman in tech that is even worse than as a woman in tech.

Dalia's account reveals how gender- and race-based discrimination shapes status hierarchies and divisions among tech workers (see also Wu 2020). While women have carved out space in certain tech professions like UX design, these fields not only have "glass ceilings" for women designers but also carry less status than male-dominated tech professions, such as software development or data science. It can be argued that these more dominant professions secure their superior status through forms of "techno-masculinity" (Poster 2013). Along these lines, this book reveals how new forms of masculine behavior are performed in a sector where women, as Hicks (2017) has demonstrated, have been systemically excluded, ever since their displacement from the central technical roles they had occupied during the rise of the industry in the 1940s. Furthermore, Dalia's experience shows how these patterns of marginalization run even deeper for people of color. Beyond facing sexism, she has struggled with racism at both interactional and institutional levels, including through subjugating visa regimes. Many researchers and public figures have problematized a lack of diversity in the tech industry and scrutinized how this results in the development of biased technologies (Benjamin 2019). This book offers a different perspective by considering the sociocultural processes that perpetuate the underrepresentation of marginalized social groups within the tech industry.

However, while my research documents these systems of differential inclusion, this book emphasizes the social codes cultivated by tech workers that I found to be shared across these internal differences and divisions. While she experiences distinct forms of marginalization, Dalia shares key aspects of the subjectivity of both Rachel and David: She is also on a quest to find middle-class success and moral worth in the tech industry. Dalia's first job as a financial consultant left her frustrated that she could not pursue these twin interests, ultimately leading her to a new professional orientation: "I thought it would be more conducive to actually build something meaningful with the client that actually addresses their problem." Drawing implicit symbolic boundaries vis-à-vis finance and consulting, Dalia narrates her entry into tech through the themes of reflexivity and social impact. Integrating social impact in her role is essential for her—something she found impossible in finance or consulting. Just like Rachel and David, though, she translates this ethos of impact into one of reformism:

> I need to align my designs with stakeholder needs. So, regarding the business needs, I mean, I can create the most amazing user experience for a user but if we can't generate profits, or if we can generate some sort of ROI for this, it's futile for the company, right? We'll just go bankrupt, essentially.

Dalia positions herself as a mediator, oriented toward brokering between different parties with different needs. Her identity is not that of an entrepreneurial self, but neither would her ethos satisfy classical Marxist criteria of a revolutionary subject. Dalia is not willing to risk the success she has achieved against a lot of odds, though she also does not want to sell her soul. She is hopeful but not utopian, critical but not radical. Her core aspiration is to harmonize economic and moral imperatives. My empirical research suggests that despite the internal divisions between tech workers, this deeper codex binds them together. The ethos of brokering and balancing conflictive needs, instead of embracing purely market-driven orientations, differentiates tech workers from other rising upper-middle-class professionals, such as those in finance but also consulting or accounting (Lupu and Empson 2015; Neckel et al. 2018; Neely 2022; Spence and Carter 2014).

Grasping and coming to terms with the distinctive social codes of tech workers is vital for anybody interested in the nexus of technology and society. As Fourcade and Healy (2024, 100) argue, "Our machines classify people because we do." If we understand technological development as a process shaped by social forces (Noble 1984; Ramtin 1991), then tech workers can be considered the architects of digital space and its various "classification situations" (Fourcade and Healy 2013). They play a pivotal role in shaping life in our digitalized societies through their decisions around the configuration and design of technologies, encoding their worldviews and values into the digital tools we rely on every day—what science and technology scholars call "inscription" (Callon 2024). If we hope to regulate artificial intelligence, tackle discriminatory algorithms, or redesign social media platforms, we must engage with and decode the multifaceted identities of the tech workers who engineer them. The *inscription power* of these professionals is essential to any efforts of recalibrating our digital tools toward the public good. Visions of tackling discriminatory algorithms and alienating apps, establishing new models of "platform cooperativism" (Grohmann 2023; Scholz 2016), or propelling something like digital "post-capitalism" (Mason 2016), will remain utopian without the support of significant numbers of tech workers.

The codes of tech workers also matter for a better understanding of the contemporary class system. Those operating the professional backstage of digital capitalism occupy a pivotal middle ground between the entrepreneurial class at the top of the tech world and the class of highly precarious digital laborers at the bottom. By studying the identifications and social codes of tech workers, we can map out what potential class alliances exist in the digital economy. Due to their middle-layered position, tech workers have the potential to build strategic class coalitions with other groups of digital workers, such as gig and crowd workers. The subjective orientations of digital professionals also have ramifications beyond the digital economy, as actors from other sectors observe and mimetically adapt to structures of the tech industry (Gerbaudo 2018, 66). Non-digital companies and employees have been shown to increasingly mimic the work forms and organizational cultures of Big Tech start-ups and enterprises (Chen 2022, 3). As Tarnoff and Weigel (2020) have argued, nowadays every company wants to be a tech company. Hence, tech workers are at the heart of an industry that is highly influential not only technologically but also symbolically.

With this in mind, this book demonstrates that tech workers cultivate new forms of identity that translate into the making of a contradictory middle-class fraction. To understand the particularities of this emerging social formation, I argue that we need to revise our established concepts for analyzing knowledge workers. Drawing on the existing literature that describes knowledge workers as "entrepreneurial selves," I had expected tech workers to align relatively closely with this neoliberal ideal (Bröckling 2015; Brown 2015). At the outset of my study, I imagined I would meet a new breed of market-fixated actors. However, nearly every conversation compelled me to revise these assumptions. It became clear to me that the ideological force of neoliberalism has lost steam within this rising class of professionals. Not only do tech workers' hearts and minds deviate from the ideal type of the entrepreneurial self, but this middle class in the making actively seeks to distinguish itself from the social figure of a market-oriented self. As we will see, tech workers' cultures and morals are, in part, a response to broader societal crises that have also shaped other segments of society. Several of their codes, shared with other groups, provide insights into shifting middle-class ways of life in a general sense. However, what sets tech workers apart as a distinct middle-class fraction is their unusual stance that involves pronounced social critique and growing disillusionment with neoliberal ideals even as they have reaped substantial rewards from recent economic and technological shifts. Furthermore, it is important to note that this study postulates a complex, open-ended transformation rather than a consolidated break with neoliberal principles. I stress the double-edged ambivalence of many seeming renunciations of the entrepreneurial ethos, and the ongoing, contested process of class formation.

The realities of the tribe of tech workers I came to know forced me to reconsider my theoretical concepts. Instead of finding heightened market orientations, I was confronted with the emergence of a *post-entrepreneurial self*. The chapters that follow offer a rich account of the underlying subjectivation processes that undergird this self. It will become clear throughout that the engineers and designers of our digital world are not merely a passive entourage in thrall to the visions of tech entrepreneurs. Recognizing their agency and their worldviews is the first step to mutual understanding, potentially enabling collaborations and coalitions between this class fraction and various stakeholders, including other groups of workers, unions, political parties, community initiatives, NGOs, educational institutions, and informed citizens. At the same time, this study does not present a naïve celebration of tech workers as progressive actors. Reading their codes in the context of theories of power surfaces many ambivalences, including covert strategies and a tendency to retreat to individualist responses to structural issues.

My research reveals tech workers as a contradictory class in the making, as they cultivate both a rejection of and a latent alignment with neoliberal values. Furthermore, when we consider the emergent post-entrepreneurial subjectivity of tech workers in the context of recent discursive and structural transformations, it becomes clear that a certain institutional appropriation of their social codes has been underway. That is, many of their emancipatory dispositions are tacitly co-opted to fuel a moral rejuvenation of capitalism. As we will see, capitalism and resistance to capitalism form no real antinomy (Boltanski and Chiapello 2018; Rider 2021). In mapping out the *contradictory class identity* of tech workers, who find themselves at the crossroads between a spirit of emancipation and yet another spirit of capitalism, this book offers important insights into contemporary professional identity and class formation and illuminates the conflicted and fragmented position of those who create the digital devices and services reshaping our world.

1 INTRODUCTION

In the public imagination, the latest technological advancements are usually credited to members of the entrepreneurial class. The popularity of ChatGPT is attributed to Sam Altman, the dominance of Amazon to Jeff Bezos, and Apple's ascent to Steve Jobs. The social making of digital transformations, however, is a process that hinges on an army of workers. Just as Rome wasn't built in a day, new digital technologies are never the product of a single individual. Platforms, devices, apps, and AI applications are the products of countless people operating within complex supply chains. The story of digital transformations is also a story of so-called tech workers, the highly skilled professionals who operate in the shadows of the entrepreneurial class. Without their labor and expertise, the steady stream of "innovations" digital capitalism depends on would quickly grind to a halt.

Despite their critical role, little is known about the population of professionals behind the scenes of digital capitalism. Tech workers rarely surface as a group of interest in the social scientific literature concerned with digital labor. Though tech workers' expertise is highly sought after and their technological influence significant, the digital labor debate has concentrated on other groupings—in particular, the highly precarious workers that have emerged from within the digital economy, such as gig and crowd workers (Graham et al. 2017; Lehdonvirta 2018; Wood et al. 2019). While this book builds on the few pioneering studies on contemporary tech workers available (e.g., Amrute 2016; Irani 2019; Marwick 2013; Neff 2012), it's safe to say that digital labor scholarship as a whole focuses less on the middle layers of the digital economy than on the emergence of a cyber proletariat (Dyer-Witheford 2015; Huws 2001) or new digital underclass (Dawson 2002; Gray and Suri 2019). One aim of this book is to complement the crucial insights yielded by these strands of research, to offer a more relational understanding of labor and class in digital capitalism.

In spite of recent layoffs at Big Tech firms, tech workers continue to constitute a substantial workforce in the contemporary economy. Digital hubs exist all around the world. Industrial firms are trying to metamorphose into tech firms. Bullish startups with the aspiration to become unicorns continue to emerge and are hiring. Even Big Tech firms are growing their workforce in certain areas. *The Economist* (2024) reports that between 2019 and 2022, Amazon, Apple, Meta, and Microsoft added roughly two thousand new AI professionals per year. While gig workers often travel the streets in their work, tech workers occupy trendy coffee shops, coworking spaces, and the glass-windowed offices of start-ups and established tech companies alike. Their labor power is required for analyzing large datasets, programming software and algorithms, developing machine learning and AI operations, and designing the user interfaces of the latest apps. As high earners, most tech workers can still afford (albeit with increasing difficulty) the skyrocketing housing and rental prices of San Francisco and Austin, Berlin and Dublin, Mumbai and Nairobi. The tech workers I interviewed can be considered members of the upper middle class. My interviewees had yearly salaries of around $162,000 in the US and $85,000 (73,000 euros) in Germany (different living costs will apply), in addition to high volumes of cultural capital.[1] With their sought-after expertise and elevated status position in the world of work, tech workers fit into the category of "professionals" (Abbott 1988; Muzio et al. 2011), though, interestingly, they typically prefer to self-classify as "workers."

While some protest tech workers as harbingers of gentrification, others welcome and celebrate their potential to rejuvenate the promise of economic opportunity. This latter camp includes universities across the world who increasingly offer specialized courses and training for aspiring tech workers. Countless Instagram and TikTok channels feature discussions of career opportunities in tech, often with the implicit message that a secure place in the shrinking layers of the (upper) middle class in still in reach. In the political arena, governments compete to become "headquarter towns." Cities and nations seek to attract tech companies and the high-paid employment they bring, often undercutting others with tax breaks and other incentives. The latest developments around generative AI only intensify this competition. With profits stagnating in many economic industries while tech companies continue to dominate the lists of most-valuable companies, the tech industry has become the "crown jewel" of contemporary capitalism (Tarnoff 2020).

The tech industry's rise to dominance raises crucial questions about the people who comprise its professional class. What kind of subjectivity characterizes the engineers building the tools and platforms we use daily? How does this coding class, positioned in between the entrepreneurial digital capitalists and the highly precarious digital laborers, negotiate its middle-layered class location? Given that the bulk of researchers have focused on highly precarious digital laborers, we know surprisingly little about tech workers' identities and values. Nonetheless, a small number of significant social scientific studies provide valuable information, though some rely on data that dates back several years. This book builds on their insights while aiming to shed light on the latest generation of tech workers.

Gina Neff's book *Venture Labor* (2012) provides an insightful account of the early phase of digital capitalism. Neff's central claim is that the rise of the New Economy in the late 1990s corresponded with its employees increasingly adopting a positive framing of risk. Rather than regarding short-term contracts and uncertain career paths as threats, new media workers envisioned them as opportunities, leading tech workers to blame themselves when they lost jobs in the dot-com crash. Tracing the contours of an entrepreneurial subjectivity among tech workers, Neff connects their micro- and meso-level codes with broader macro-level code shifts, including the dominant neoliberal political rhetoric of the Reagan and Thatcher era (Neff 2012, 46). Bergvall-Kåreborn and Howcroft (2013) reach a similar conclusion in their analysis of mobile app developers and design professionals, diagnosing the subjectivities of tech workers as characterized by self-control and self-rationalization. Along similar lines, Marwick (2013) argues that US tech workers frequently engage in practices of self-commercialization to fit the widespread ideal of micro-celebrity status. My study is in dialogue with this literature while revealing a different pattern: the shape of a new middle-class identity among contemporary tech workers.[2] In the context of multiple crises, I found that the majority of my interviewees cultivated a transformation of the ideal of the entrepreneurial self by departing, in a contradictory fashion, from a holistic market orientation.

Another line of research analyzes the corporate workings of the global tech industry in terms of the divisions and differences sustained within the field. Sareeta Amrute (2016, 2019, 60) shows how the subjectivities of Indian tech workers laboring within Western digital economies are deeply affected by processes of racialization and class. She demonstrates how Indian tech workers using temporary work visas in Germany are at once racially marginalized and upwardly mobile, their middle-class aspirations simultaneously propelled and stifled. Amrute's findings illustrate the ways in which the contemporary global economy is structured according to racialized mechanisms of "differential inclusion" (Mezzadra and Neilson 2013) that leave tech workers exploitable to different degrees and thereby fracture any potential solidarity among them (Amrute 2016). Another key study of the operations of power and social inequalities among tech workers comes from Lilly Irani. In her analysis of tech workers laboring in the Global South, Irani argues that the young and well-educated Indians who pursue careers in the national tech economy show strong motivation, goodwill, and the ambition to contribute to a "better" Indian society, but she points out that this desire is subsumed under economic logics (Irani 2019, 141). By demonstrating how workers' sense of hope and national pride are ultimately used to propel more flexible and precarious work relations within an Indian tech hub, Irani generates crucial insights into habitat-specific forms of subjectivation. Building on this strand of research (see also Carrigan 2024; Gupta 2019; Radhakrishnan 2011), my work likewise analyzes divisions within the heterogenous workforce of tech workers, addressing differences of gender, race, and nationality. At the same time, this book centers on the *shared social codes* that tech workers in the Global North develop *even across these internal disparities and divisions*. My inquiry sheds light on the cultural and moral ties that unite the ensemble of actors operating backstage in the tech industry and distinguish them as a digital middle class.

A third central strand of research on tech workers focuses on their activism and emergent collective organization. This strand is shaped by the pioneering work of Ben Tarnoff (2020), who diagnosed an emergent "tech worker movement." Tarnoff points to a structural shift in an industry that was long considered an anti-world to activism and unions, citing examples of tech worker walkouts over workplace discrimination, protests against developing artificial intelligence targeting software for military drones, and emerging unionization efforts. Relatedly, Nedzhvetskaya and Tan (2024a) discuss how AI tech workers have problematized recent technological developments, while Niebler (2023a) analyzes the ways tech workers build strategic alliances with low-paid gig workers. These studies highlight increasing tensions and power struggles in tech but also, somewhat reluctantly, remind us that collective action and unionization efforts among tech workers are not mainstream. According to Tan, Nedzhvetskaya, and Mazo (2023), media reports of tech worker activism have plateaued or dropped since 2019, and Niebler's inquiry focuses on just a handful of instances of cross-class alliances. If there is a tech worker movement, it does not appear to be a mass movement. This situation has led some activists to claim that "there is something missing from tech worker organizing" (Molinari 2020). However, to date, we lack a clear understanding of why the emergent tech worker movement has been unable to gain more traction. To tackle this question, my study homes in on the hearts and minds of tech workers. Driven by an interest in how ordinary tech workers in the US and Germany see the world and understand themselves, I engaged in conversations with tech workers recruited across different channels and irrespective of the question of whether they have engaged in collective action or not.

My study also complements the nascent literature on tech workers by providing comparative insights across national settings. Existing studies are usually limited to one national field site and rarely analyze the phenomenon of tech work across nation-states.[3] I examine how tech workers in both the US and Germany cultivate new social codes amid multiple contemporary crises. Across these contexts, I explore how several fundamental crises of the neoliberal conjuncture have deeply impacted the subjectivity of tech workers in recent years. First, there is a political crisis over legitimate representation, which has especially influenced US tech workers both through debates about bias in digital technologies as well as the 2016 Cambridge Analytica scandal that shocked the tech industry by revealing its role in Donald Trump's presidential victory. There is also an economic crisis of widening inequality. Tech workers are confronted with this crisis not only at work, where their superiors often rank among the world's wealthiest individuals, but also at home as they often live in cities with unprecedented levels of homelessness and overheated housing markets, where buying property stretches even their own comfortable salaries. Finally, the ecological crisis of climate change and the Covid-19 pandemic have sparked widespread questioning of our liberal economic markets and individualist societies. While my study takes account of the important differences across nations, I point to a cross-national disillusionment with neoliberalism among tech workers. This disruptive climate has led tech workers in both the US and Germany to pivot away from market orientations, though to different degrees and in contradictory ways. Crises can challenge even the most sedimented social orders and codes (Beck 1992).

NEOLIBERALISM AND THE ENTREPRENEURIAL SELF

To ground my study, let me sketch some historical backdrop and key theoretical frameworks. The epoch of neoliberal capitalism emerged in the 1970s, when a more globalized, networked, and deregulated capitalist regime began to take hold in the Global North (Bell 1973). The material roots for this transformation can be traced back to economic stagnation and advancements in logistics as well as information and communication technologies, which enabled widespread outsourcing and on-demand economic production (Reckwitz 2020, 81). In parallel, government safety nets were cut down, Keynesian beliefs in a strong state faded, and faith in the market as an organizing principle gained popularity even beyond the economic field (Mau 2015). A neoliberal code of conduct valorizing anti-government sentiments and individual and collective market orientation gained hegemony (Foucault 2008). The process of neoliberalization cannot be adequately grasped without taking into account the shifting cultures of subjectivation, whereby new forms of subjectivity arose in parallel to economic, political, organizational, and technological transformations. The notion of the *entrepreneurial self* (Bröckling 2015; Rose 1989; Skeggs 2004) sums up manifold elements of the new ideal subjectivity (see also Foucault 2008). This new self refers to an employee who behaves like an entrepreneur in many ways and in many spheres of life. This is a self that prefers networks over hierarchies and longs for autonomy while remaining always available to and for work (Pongratz and Voss 2003; Sennett 2007). This self is less attracted to unions and feels more at home working on temporary projects than in life-long careers (Boltanski and Chiapello 2018, 90). Unlike the *organizational self* (Gregg 2012; Whyte 1956), considered to be the ideal-typical subjectivity of industrial capitalism, the entrepreneurial self does not want to stay within one company for a lifetime and dislikes the idea of being a cog in a corporate machine. This self is on a journey to find meaning, singularity, and creativity within work (Boltanski and Chiapello 2018; Florida 2002; Reckwitz 2020) and considers the strict division between work and life spheres unnatural: Nine-to-five jobs come to seem like prison sentences. There is a stark contrast, here, between the organizational self, anchored in values of stability, collectivity, loyalty, and conformity, and this post-Fordist entrepreneurial self.

These transformations were mutually dependent. As capitalism shifted toward neoliberalism, cultures of subjectivation increasingly devalued the organizational ethic and middle-class employees self-classified as project-oriented, creativity-seeking, and self-rationalized individuals. A Faustian pact was born. Paradoxically, many of the "virtues" of the entrepreneurial self originally emerged as a critique of capitalism (Boltanski and Chiapello 2018). Anti-capitalist counterculturalists, artists, and scholars had long condemned industrial capitalism as a world of work that circumscribed creativity and autonomy. Yet their "artistic critique" of alienating work relations unintentionally contributed to a rejuvenation of capitalism (Boltanski and Chiapello 2018, 343). Their critique, which foregrounded the desire for autonomy, was appropriated by management gurus, employers, and neoliberal politicians to establish a more flexible and project-structured world of work whose flip side was shorter employment contracts, de-unionization, atomization, and self-exploitation.

Of course, this broad-stroke recapitulation cannot capture the nuances of several decades of work, culture, and power relations, and it is important to emphasize that the *organizational self* and the *entrepreneurial self* are ideal types, theoretical constructs that do not exist empirically in pure form but rather accentuate patterns in social conduct (Weber 1949, 90). Furthermore, throughout the history of capitalism, it has been possible for individuals to embody the effects of both types simultaneously (Reckwitz 2010). Nonetheless, the outlined framework establishes a changing tide in work and subjectivity formations. The ideal type of the entrepreneurial self provides us with a sense of the dominant middle-class culture of subjectivation that preceded and undergirded the rise of digital capitalism. The concept serves as a valuable reference point and benchmark for analyzing the identity formation of tech workers, allowing us to identify new characteristics and to contribute to larger social scientific debates around the subjectivity of white-collar workers and professionals in contemporary capitalism.

SUBJECTIVITY AND BOUNDARIES

This research participates in a rich tradition of examining capitalist dynamics through the lens of historically specific values and beliefs (Boltanski and Chiapello 2018; Thompson 1966; Weber 2005). Economic action is understood as sociocultural action, driven by schemas of perception, recognition, and action. Put bluntly, I am interested in the motives, beyond earning money, that get people out of bed and engaged in work that sustains capitalism. Furthermore, my work sheds light on the cultures and knowledge orders that shape technological developments as well as market structures (Akrich 1992; Callon 1998; MacKenzie and Wajcman 1999). Drawing on science and technology studies, I seek to get to grips with the social characters behind the data gaze, the professional subjects who play a key part in the birth of the platform economy. Though my study does not directly examine the practices of tech workers, I provide insights into the deep-layered social codes that govern the technical coders and designers who are professionally shaping the algorithms, apps, and platforms that increasingly define our societies.

Foucault's work provides a theoretical foundation for my understanding of subjectivity as the historically specific ways human beings experience themselves in the context of discourse and technologies of the self (Foucault 1982, 2012a). This approach replaces anthropological constants with a conceptualization of subjects as the genealogical outcome of self-dynamic knowledge orders, with selves holding the capacity to critically reflect on these and transform themselves (Foucault 2002, 2012b). In less abstract terms, Foucault helps us to understand that the kind of identity human beings develop takes place in historically specific contexts shaped by powerful institutionalized ways of thinking and acting. Furthermore, while classical sociological theories tended to conceptualize identity as something stable and coherent, Foucault's work advanced a theoretical tradition that is attentive to the fragmentation and shifts of the self. Theories of identity, however, have increasingly incorporated these insights (e.g. Hall and Du Gay 1996), blurring the boundary between identity and subjectivity—hence my decision to use the terms interchangeably in this project.

Importantly, my work will complement the concepts of subjectivity and identity with class theory. For Bourdieu (1984), the habituated schemas of perception, recognition, and action that characterize subjects are linked to class through unevenly distributed forms of capital. Thus, socialization processes not only depend on a historical context with powerful knowledge orders but also on the unequal, class-based distribution of resources, a consideration that is not given a systematic role in Foucault's account. Bourdieu's class theory also makes it possible to grasp class positions and struggles not only in terms of economic capital but also with reference to the distribution of cultural and social capital (Bourdieu 1985). Consider the early phase of capitalism, for instance, where the bourgeois class developed not just through economic force but also by culturally distinguishing themselves from the figures of the decadent aristocrat above and the uncivilized proletarian below (Skeggs 2004, 4). Only by enacting symbolic boundaries could this class of small entrepreneurs and merchants culturally and morally secure and legitimize their economic positioning and, ultimately, come to dominate the aristocratic class. In combination, the relational and power-oriented frameworks offered by Foucault and Bourdieu allow me to examine how subjects (or selves) are brought into existence in the context of different systems of knowledge and different forms of capital.[4]

Though the theories of Foucault and Bourdieu form a central part of my theoretical tool kit, I supplement them with other theoretical concepts in this work, such as Lamont's notion of symbolic boundaries (1992, 2002). This heuristic allows us to differentiate between socioeconomic, cultural, and moral processes of distinction (Lamont 1992, 4). Importantly, from a sociological perspective, moral distinctions are neither more or less virtuous than socioeconomic or cultural ones, and this study does not seek to judge subjects' moral boundaries but rather to examine them as a key mechanism through which group closure processes occur. Another central concept I employ throughout is the notion of *social codes*. While the term "code" is currently mainly understood as a technical computing term, I seek to adapt and reconfigure this vocabulary from a social scientific perspective, pointing toward the field-specific rules and normative horizons that underlie processes of subjectivation. Taking the form of explicit, and especially implicit, rules, social codes amount to frameworks that strongly shape the behavior and identity of individuals in their respective social fields (Anderson 2000; Benjamin 2019). At the same time, social codes are also always objects of struggle since they regulate the flow of different forms of capital (Pistor 2019).[5] Furthermore, social codes entail a *normative dimension*. Without a sense of what is good or bad, right or wrong, the self would decompose and thus be unable to navigate a social field in any coherent way (Butler 2005; Raza 2022; Taylor 1992).[6] Subjects always cultivate a code of conduct: Selfhood is intertwined with notions of the good (Durkheim 1997). However, while Durkheim proposes normativity as the basis for societal solidarity—the glue that keeps society together—I also inquire into the ways in which normative codes are tied to the making of conflictive groupings (Lamont and Molnár 2002).

METHODOLOGICAL APPROACH

Building on this heuristic framework, this book is based on original qualitative empirical research aimed at analyzing tech workers in terms of their subjectivity and class formation. In line with their self-understandings, tech workers can be defined as the ensemble of trained and mostly high-paid knowledge workers within the tech industry. Thus, when deploying the terms "tech worker" and "tech company," I am not simply referring to any workers or companies whose operation is heavily reliant on technologies. Instead, my research homes in on those actors who are embedded in the contemporary "tech industry" or "digital economy," understood as the relatively autonomous economic field that arose hand in hand with advancements in information and communications technologies and whose analytical core is composed of economic actors linked to the commercial internet (Neely et al. 2023; Ziegler 2022). The tech industry can thus be understood as a social microcosm that is distinct on the interrelated grounds of its specific economic activities as well as its work cultures. Importantly, this industry appears to be growing with more and more companies changing and adapting their business activities in response to digital and cultural transformations.[7]

My empirical analysis focuses on tech workers in the US and Germany. While the US dominates the global tech industry, Germany is considered leading in Europe's tech landscape. Furthermore, the two countries represent contrasting market economies of the Global North: The US is a liberal market economy, while Germany can be considered a coordinated market economy (Hall and Soskice 2001). Conducting my research at these two national sites allowed me to gain comparative insights. Studying both locations revealed pronounced shared patterns that, given their different contexts, allow me to make tentative arguments about tech workers across the Global North and contribute to social scientific understandings of tech work as a transnational phenomenon. At the same time, it is important to acknowledge that this study is ultimately limited, in terms of habitats, to postindustrial societies in the Global North, and my work does not capture the significant work and roles of tech workers beyond this geographic region

(Beltran 2018; Irani 2019; Seto 2025; Takhteyev 2012). Nevertheless, since the professional workforce in tech companies in the Global North is highly international, I interviewed several tech workers who had migrated to the US or Germany from the Global South, and I discuss their distinct experiences as tech workers especially in chapters 3, 5, and 7, seeking to analyze the substantial connections between the Global South and the tech workforce in Germany and the US.

Guided by the principles of grounded theory and abductive research (Glaser and Strauss 1967; Timmermans and Tavory 2012), I combine qualitative interviews and discourse analysis to explore the social codes of tech workers across the US and Germany (a more detailed account of my analytical strategy, data collection, and analysis as well as a discussion of my positionality is contained in the appendix). In light of the vast number of tech worker professions, I pursued a strategy of maximum contrast sampling, focusing on two groups with distinct positions within the professional field of tech work: UX designers and data scientists. UX designers primarily work "front end" and operate mostly with qualitative data. Their work is considered front end because users directly engage with their work products: UX designers are hired to enhance the usability, accessibility, and enjoyment provided by digital products and their interfaces. Data scientists tend to work at the "back end" and deal with quantitative data. Users engage with their work indirectly, which is why it is denominated "back end." Data scientists are responsible for the collection and analysis of large datasets as well as the development of data-based algorithms. While a UX designer at a media streaming platform may be responsible for designing the way in which movies appear to customers on the homepage, a data scientist will develop the algorithms that determine which movies are suggested to the viewer. Given the different work tasks and organizational roles, I hold that similarities between these two distinct professional groups provide insights into more general characteristics of working in tech.

Between 2020 and 2022, I recruited interview partners through diverse channels. I first made contact and was able to interview a number of UX designers and data scientists at both start-ups and established tech companies through personal connections at both field sites.[8] In Germany, I established connections primarily through joining a coworking space in Berlin. I visited the space, located in the central district Mitte, for several months in 2020 and 2021, enabling me to get in touch with a range of individuals who were either tech workers themselves or could refer me to people who were. In the US, I managed to get in touch with a number of tech workers through contacts I had established during an academic stay at Duke University (2018–2019), where I took a course in the newly established "Interdisciplinary Data Science" master's program. This led to direct contact with several (ongoing) tech workers, and the academic staff involved in the program kindly connected me with an array of employees at tech companies in Silicon Valley and beyond. It is also worth mentioning that it was this course, which I originally enrolled in to learn methods for quantitative textual analysis, that initiated my interest in tech workers as a sociological object of study.

In addition to recruiting interview partners through personal contacts, I deployed a snowball strategy by asking my interviewees to provide me with further contacts, activating my social capital. I also contacted tech workers outside of my personal network through the online platform LinkedIn. Finally, I made use of the online community website of the Berlin coworking space I had joined. Their directory allowed me to get in touch with more tech workers in Germany as well as some who had moved to the US. These four methods—personal contacts, snowball sampling, LinkedIn, and the coworking space's online directory—did not result in a representative sample. However, they provided me with diverse online and offline entry points to tech workers, helping to mitigate potential selection biases and offering a more varied perspective on the field.

I subsequently conducted fifty-two semi-structured interviews with data scientists and UX designers. I focused primarily on recruiting tech workers at the mid-level career stage. This means I undersampled both tech workers who very recently entered the field as well as more senior tech workers in management roles. In terms of age, my interviewees were between twenty-five and fifty-three years old, with a mean age of thirty-four. Of my twenty-six US participants, twelve were foreign nationals (Asia was an especially common continent of origin, followed by South America and Europe). In Germany, sixteen of my twenty-six interviewees reported an international background (mostly European, with some from South America or Asia). In terms of gender, seventeen of my interviewees identified as female, one as non-binary, and thirty-four as male. This gender makeup roughly corresponds with the statistics of publicly available diversity reports by the Big Tech companies Facebook, Google, and Microsoft, where women occupied around 25 percent of tech roles in 2020. In terms of race, my sample also indicates an unequal representation of marginalized social groups: Thirty-six of my interviewees identified as white, seven as Asian, six as Latino or Hispanic, two as African American or Black, and one as Arab. Again, these proportions roughly correspond with the reports of tech companies themselves, with white and Asian tech workers occupying around 90 percent of the "technical roles" at Facebook, Google, and Microsoft in 2020 (notably, official reports suggest that I undersampled individuals from Asian backgrounds). Turning to class, I noticed that the corporate diversity reports generally do not inquire about this element of the social backgrounds of their employees. My explorative study, however, indicates that a severe underrepresentation is also at play here. On the basis of the parental occupations of my interviewees—which showed a high frequency of professions such as teachers, professors, engineers, lawyers, and doctors—I interpret that roughly 80 percent of the individuals in my sample have either a middle- or upper-middle-class background. This suggests that the tech industry is a habitat of middle-class reproduction rather than a corridor for upward mobility for the working class.

In addition to these interviews, I also conducted a discourse analysis in order to examine the interpellation of tech workers. Discourses exert power by creating identity offerings (Foucault 2002). I analyzed study programs (fifty documents) and job ads (138 documents) as a means of contrasting tech workers' self-codifications, as reconstructed in the interviews, with the codification of their subjectivity in institutional discourses. In other words, having analyzed how tech workers understand themselves, I turned to examine how universities and tech firms imagine and summon them. While this approach produced a rich triangulation of different empirical datasets, it is important to acknowledge that this study does not offer an exhaustive understanding of the social phenomenon of tech workers. Although the type and amount of data collected here allows my book to offer theoretically saturated answers to my primary research questions, I do not claim to offer an encompassing sociological account of tech workers, which would require significantly more data as well as further methods of study. This study aims to provide a systematic exploration of tech workers' subjectivities but not their everyday practices. Ethnographic research would be needed for a full perspective on the (possible lack of) translation between subjective orientations and practices. Furthermore, observational studies of tech workers' practices could generate thick accounts of their bodily entanglements with an array of relatively self-dynamic material objects, thereby bringing an analysis of subjectivity, identity, and habitus into contact with other research paradigms, particularly actor-network-theory (Callon 1984; Latour 2005). Quantitative studies, on the other hand, would be required to gain representative accounts of tech workers' socio-structural characteristics and backgrounds.

CONTOURS OF A POST-ENTREPRENEURIAL SELF

Despite the moderate size of my dataset, I was able to identify a number of recurring sociocultural schemas among tech workers. Based on my explorative research sample, my study describes the contours of a new subjectivity that signals a transformation of the entrepreneurial self without indicating a wholesale departure from neoliberal rationality. While notable differences across national sites and among professionals exist, I found widely shared codes across worldviews, professional roles, and lifestyles.

With regard to worldviews, it is possible to diagnose a widespread *return of social critique* that indicates, in turn, a new moral figuration. Boltanski and Chiapello (2018) distinguish between the "social critique" of capitalism, which problematizes inequality and exploitation, and the "artistic critique," which takes aim at a perceived lack of autonomy and creativity within capitalism. People who articulate artistic critiques resent capitalism primarily because it leads to homogeneity and alienation. While this can be considered the dominant critique in neoliberal societies (Boltanski and Chiapello 2018, 169), my research suggests that the social critique is regaining momentum. In contrast to the neoliberal self, whose conception of the good and right centers on expressive individualism and a strong market orientation, today's tech workers are more sympathetic to collectives and articulate their critical capacities by problematizing economic inequality and a lack of diversity. They rarely express dissatisfaction regarding the realization of autonomy or creativity in their work, but instead express sympathy for unions, value open-source activities, and problematize the precarious social situation of gig workers. Furthermore, tech workers are concerned with the underrepresentation of marginalized social groups and the possibility of encoding "bias" in digital technologies. Their investment in issues of diversity extends the classic social critique. Through reflective data work, many tech workers aim to establish a new rule of data and code that departs from a naïve "techno-solutionist" belief system in which tech is seen as the solution to all social issues and problems (Morozov 2013; Nachtwey and Seidl 2020).

Contexts including growing economic inequalities, overheated housing markets, political upheavals about the representation of women and ethnic minorities, the Covid-19 pandemic, and the climate crisis all undergird the return of social critique. In the context of these challenges, the good and the right became tied to a sense of fairness and justice wherein subjects do not seek only to avoid harming others but aim to actively promote others' welfare (see also Rider 2021). This other-oriented morality, in its departure from the more self-oriented normative codes typical of the entrepreneurial self, seems to me the primary impetus behind the limited but potentially significant unionization efforts and protests led by tech workers (Boag et al. 2022; Su et al. 2021, 22; Tan et al. 2023; Tarnoff 2020). At the same time, I also interpret this socially critical codex as a form of symbolic capital and a key mechanism in tech workers' status and boundary-making processes. While my research points to a genuine longing for social change that can manifest in action beyond self-presentations, most of my interviewees' articulations of social critique were not accompanied by reports of practices corresponding with this critique. Thus, while it is important to identify the transformation of the entrepreneurial ideal already evident in the self-presentations of tech workers, and consider the space of possibilities this opens up, we must also consider the extent to which these displays of caring about equality and social justice may constitute new strategies (without strategists) of distinction. Furthermore, this book also critically examines the latent forms of discrimination and marginalization that prevail in the tech industry despite its widespread claims to valorize diversity. While the social figure of the "tech bro" (Lobo 2014) has been largely delegitimized not just within the public but also among tech workers themselves, I suggest that new forms of more latent "techno-masculinity" (Poster 2013) are emerging within the industry.

Regarding professional roles, this book identifies among tech workers a prevalent logic of hybridity. This codex offers insights into how tech workers align their social critique with their professional role and their organizational embeddedness. My research highlights the importance of the code of *brokering economic and societal needs* in understanding how tech workers manage to integrate moral and economic action. That is to say, the professional ethic of tech workers centers on balancing different requirements. By aiming to find the "sweet spot" between business needs and the needs of the users of platforms, tech workers aim to avoid a rupture of their identity—a "habitus clivé" (Bourdieu 2008; Friedman 2016). The moral codex behind this hybrid professional role stems from a deep longing to be simultaneously *middle-class wealthy and morally worthy*. The typical tech worker I encountered aspires to both maintain a middle-class standard of living and also to achieve moral worthiness in the other-oriented sense of improving people's lives.[9]

Tech workers also manifest a hybrid social code in a different sense: by synthesizing the figure of "the nerd" and the figure of "the communicator." Tech workers cultivate an attitude of *post-nerdiness* at work—they value a combination of technical skills and communicative and empathic capabilities. This hybrid codification drives internal hierarchies and allows tech workers to draw boundaries vis-à-vis supposedly more one-dimensional professionals, such as statisticians (relative to data scientists) or nondigital designers (relative to UX designers). A final form of hybridity manifests in a code of *inclusive domination*, expressed differently in terms of educational barriers and racialized relations of inequality. With regard to the former, I found that tech workers valorize nonspecialized academic programs, indicating credibility tactics that paradoxically employ inclusionary logic to establish occupational boundaries. This contrasts with the reliance on exclusive educational entry barriers seen in traditional professions (Abbott 1988).

However, other formal entry barriers do exist. Specifically, applying Bhabha's (1994) postcolonial theory of hybridity allows us to identify the specific forms of inclusive domination that confront Global South tech workers. As I will later discuss, migrants from the Global South must balance experiences of being simultaneously inside and outside of the Global North's tech industries due to subjugating visa regimes and precarious forms of belongingness.

My research also identified three defining lifestyle codes among tech workers: *comfortable exploring*, *ordinariness*, and *mindfulness*. As regards the first, I found that tech workers strive to explore new sites, situations, and identities, but within the bounds of economic and moral comfort. They like to live on the edge without risking full exposure. They accept the demands of a hyper-globalizing and dynamic social order under the condition of economic and moral stability. The lifestyle of ordinariness, on the other hand, manifests in a foregrounding of inclusive, rather than exclusive, tastes and leisure activities. This is surprising given the mostly upper-middle-class salaries of tech workers and provides new insights given that cultural sociology has long associated high status with highbrow cultural distinction (e.g., Bourdieu 1984; Reckwitz 2020): Think of the manager who likes to golf or the creative director who tells you about their visit to an art exhibition on the weekend. While tech workers are certainly not immune to exclusive consumption habits and lifestyles, they are keen to *foreground* their investment in down-to-earth leisure activities, such as walking their dog or having a barbecue with friends. Thirdly, I also find a widespread conduct of mindfulness among tech workers. Tech workers actively protect work-life boundaries and guard themselves against burnout. Overwork carries no prestige in their culture; instead, their lifestyles prioritize the sustainable management of their psychosocial resources. As we will come to see, though, these three lifestyles manifest in ambivalent ways and generate unintended consequences.

IDEAL-TYPES, DISCOURSE, AND CLASS

The entrepreneurial self is by no means dead. There has been no radical break whereby neoliberal ideals of autonomy, flexibility, and creativity vanished from the hearts and minds of tech workers. However, only about a quarter of my interviewees showed a holistic neoliberal orientation as their dominant rationale. Ideal-typically speaking, this type can be coined *the affirmative tech worker*. Individuals who display this category in most situations are located at the orthodox pole of the professional field and typically voice little concern around the modus operandi of the tech industry and society at large. They typically hold conservative-leaning political views and often aspire to become tech entrepreneurs themselves in the future.

My research identifies two further ideal types: *the reformist tech worker* and *the radical tech worker*. The reformist tech worker type accounted for roughly half of my interviewees. This type departs from the entrepreneurial self through a heterodox cultivation of the social codes I have discussed. Reformist tech workers seek a different society and self, and regard affirmative tech workers as social climbers. Notably, however, they are not radicals: The reformist tech worker typically prefers incremental change. My interviewees David and Dalia are good examples of this type. David is concerned with ethical design and is skeptical of laissez-faire capitalism but, at the same time, told me he could imagine working for a big tech company if he felt he would be able to contribute to change from within. Dalia sees digital technologies as avenues for improving people's lives, but she is extremely frustrated with pervasive sexism and racism in the tech industry. Compared to other interviewees, Dalia is less able to resist certain market logics because her visa is linked to her work, reminding us that tech workers from the Global South who do not have access to full citizenship inhabit a smaller space of possibilities from which to translate their critique and morals into action.

The final type, the radical tech worker, performs the new social codes in a more heretical manner. This type mostly works at start-ups rather than big tech companies and seeks structural reconfigurations of economy and society. Without necessarily being full anti-capitalists, radical tech workers fiercely articulate their critique of the tech industry and they are sympathetic to alternative political concepts, such as the "socialist vision" of Bernie Sanders. My interviewee Rachel leaned toward this third type in most (not all) situations. She considers herself left-wing, identifies strongly as a worker, rather than a professional, and, unlike David and Dalia, she told me she absolutely could not imagine working for one of the Big Tech companies.

All types are ideal in the Weberian sense that individuals can resemble them more or less but will never be pure representations of one type, and my empirical analysis of the interviews will put more flesh on the bones of these three ideal types. For now, I simply want to highlight that, despite variations in nation, gender, and race, the bulk of my interviewees exhibited social codes that deviate, to different degrees and in different ways, from the neoliberal ideal of a holistically market-oriented self. Tech workers' active distancing from many foundational neoliberal attributes signals the emergence of novel cultures of subjectivation in the very heart of capitalist development that deserve serious consideration.

At this point, one might reasonably question the distinctiveness of tech workers' middle-class character. Are tech workers' post-entrepreneurial social codes not simply representative of a wider cultural shift in Western societies? This study provides a two-fold response. On the one hand, it contextualizes tech workers' subjectivity within a set of wider societal crises including the 2008 financial crisis, contemporary political conflicts over discrimination and bias, and the climate crisis. These widespread contexts remind us that social critique and new lifestyle patterns have gained prominence well beyond the professional ranks of the tech industry (Beck and Westheuser 2022; Grasmeier and Beck 2023; Piketty 2021). In this sense, this study of tech workers contributes to an understanding of a broader societal disillusionment with neoliberal principles. In particular, my key finding of an emergent post-entrepreneurial subjectivity may point to a larger trend in the middle class. Yet, at the same time, my research illuminates how tech workers put a specific spin on the return of social critique evident in their distrust of decision-making processes based on unreflective reflective data work. Moreover, studies of other professional groups tell us that social critique is not an emblematic culture among professionals in the *upper* middle class. Although tech workers' moral codes share commonalities with, for instance, academic scholars or creative workers (Brorsen Smidt et al. 2020; Patrick-Thomson and Kranert 2021; Südkamp and Dempsey 2021), similar orientations remain uncharacteristic among more affluent professionals in the contemporary economy—especially those in finance, but also those in accounting or consulting (Lupu and Empson 2015; Neckel et al. 2018; Neely 2022; Spence and Carter 2014).

Why, then, are tech workers so often assumed to embrace entrepreneurial and techno-optimist worldviews, to embody what Barbrook and Cameron (1995) coined the "Californian Ideology"? And why do several academic accounts of digital capitalism, like that of Burrell and Fourcade (2021), group tech workers together with the entrepreneurial elite as a unified social class? One explanation can be found in the fact that the institutional interpellation of knowledge workers in tech is, on the whole, more neoliberal and entrepreneurial than the subjectivity held by tech workers themselves. My discourse analysis of job ads and university study programs finds only a partial correspondence between the organizational discursive framings and tech employees' self-understandings. The way that employers and universities talk about what it means to be a tech worker differs from how tech workers present themselves. For instance, my interviewees' concerns around economic inequality, their esteem for ordinariness, and their valorization of 9–5 working hours do not find any meaningful echo in the institutional discourse. However, my discourse analysis did reveal that some tech workers' social codes are strategically appropriated in these contexts. For instance, tech firms often reframe socially critical concerns around discrimination and bias by offering an appreciation of diversity grounded in its potential to fuel innovation and increase productivity. With the reelection of Donald Trump in 2024, who received significant support from segments of the tech industry's entrepreneurial elite during his campaign and is praised by almost the entire tech elite after his successful run, it remains to be seen whether the corporate appropriation of social critique will continue. However, while the long-term trend remains uncertain, this study shows that the early 2020s were marked by a pronounced corporate focus on co-opting social justice-related issues as part of their strategic narratives. Based on my discourse analysis, I also find that there is a strategic translation of hybridity within the discourse. I reconstruct how tech workers' hybrid balancing of needs is imagined in the discourse as a harmonious process rather than a task that requires the constant effort of brokering distinct or conflictive interests. And with regard to lifestyles, I found several depictions of mindfulness rewired into the discourse as an ethic that allows for heightened work performances. The institutional discourse on tech workers seems to only align with those post-entrepreneurial self-understandings that are amenable to processes of *commodification*. Hence, while many tech workers' self-understandings reflect a deep-rooted desire for a condition beyond neoliberalism, we must also consider unintentional consequences and the possibility that this very desire can be molded into yet another spirit of capitalism (Boltanski and Chiapello 2018; Merton 1936).

It is important to emphasize here that the social codes of tech workers are not necessarily doomed to propel commodification and a cultural rejuvenation of capitalism (see also Amrute 2014, 107–108). At this critical juncture in (digital) capitalism, these codes culminate in what we might call a *double contradictory class formation*. Loosely building on the work of Wright (1976), we can understand tech workers as occupying a contradictory class location not just economically but also in their cultural, moral, and political orientations. On the one hand, tech workers' social codes exhibit emancipatory potential, demonstrating principles of reflection, (symbolic) collectivity, and solidarity. Tech workers are more than a statistical grouping of individuals with similar resources, a "class on paper" (Bourdieu 1985). Despite their internal differences and divisions, my empirical study shows that tech workers are undergoing a process of class formation based on a shared conduct of life and boundary-makings. On the other hand, though, tech workers do not constitute a fully realized class. My interviewees rarely reported engaging in collective action or workplace resistance. Despite a shared disillusionment with neoliberalism, tech workers often remain glued to individualist modes of thought. Furthermore, as I discuss later, some of the class codes exhibited by tech workers, which signal solidarity with lower classes, can actually impede social mobility from below. In line with the work of Friedman and Reeves (2020), we can read tech workers' ordinariness, for instance, as a strategy that shields this group from critique while allowing them to pull away economically. Ultimately, tech workers must be viewed as a double contradictory class fraction: They are economically positioned in between capital and labor, and their social codes are ambivalent, double-edged, and easily co-optable (see table 1.1 for an overview of tech workers' social codes).

Table 1.1
Overview of the social codes of tech workers

Worldviews	Return of social critique; concern for diversity; critical technosolutionism
Professional roles	Brokering business and user needs; cultivating post-nerdiness; combining low and high professional entry barriers
Lifestyles	Comfortable exploring; ordinariness; mindfulness
Political orientation	Sympathy for unions; preferences for left, liberal and green parties; disillusionment with capitalism within a subjective framework that is itself disillusioned with imagining other modes of economic organization; neoliberal structures are challenged from within the individualist doxa of self-reliance and self-control
Overarching subjectivation	On a quest to be middle-class wealthy and morally worthy; contradictory transformation of entrepreneurial selfhood
Culture-class nexus	Distinctive subjectivity among the ranks of upper-middle-class professions (e.g., bankers, consultants, accountants)
Institutional response to tech workers' social codes	Emergence of yet another spirit of capitalism; appropriations of tech workers' social codes via corporate humanism, play of differences, and compassionate design
Class trajectory	Contradictory class formation stemming from ambivalences, capitalist appropriations, and apparent lack of translation of moral self-understandings into corresponding practices

Coming to terms with the contradictory class identity of tech workers matters for all those interested in reshaping the digital worlds that we inhabit. Tech workers are socially relevant not only due to their growing numbers but also due to their *inscription power*: that is, the way their work shapes and structures our digital tools and worlds, including the unequal access to resources that determines life possibilities both on- and offline. Given this highly influential position, understanding and dialoguing with tech workers will be pivotal for diverse social and political projects. Unions, governments, educational institutions, activists, and other societal actors should recognize that tech workers are distinct from tech bosses and no "cogs in the machine." Rather, they are active agents with a contradictory set of economic interests and moral codes. To enlist tech workers as allies, societal actors would need to devise strategies that address these workers' immediate economic concerns but also resonate with their moral and ethical considerations. At the same time, avoiding increasing commodification and inequality will also require critical self-reflection and institutional engagement from tech workers themselves, as well as other social classes sharing similar codes.

Throughout this study, I show how the social codes of tech workers hold the potential for both domination and emancipation. While building on critical sociology and post-structuralist insights, I aim to move beyond a mode of "paranoid inquiry" (Sedgwick 2003) that focuses primarily on unmasking the hidden mechanisms of power and domination in order to consider real-world spaces of possibilities. At the same time, the account I provide does not overlook new mechanisms of exclusion and boundary-making within tech work, nor does it underestimate the adaptive capacities of capitalism. This book is written in the spirit of conceiving the social in terms of emancipation and domination at one and the same time.

OUTLINE OF THE STUDY

Chapters 2, 3, and 4 present an in-depth empirical analysis of the social codes of tech workers, with a comprehensive discussion of their worldviews, professional roles, and lifestyles based on both interviews and discourse analysis. Each of these chapters contains four subsections. The first three subsections present empirical findings from my interviews, while the fourth examines how tech workers' self-understandings align with the institutional interpellations I found in the discourse about tech workers. Chapters 5, 6, and 7 then expand into broader macro-sociological discussions of the social codes of tech workers. Chapter 5 further embeds this research within cultural class theory, systematically examining the reconstructed social codes of tech workers through Bourdieu's theoretical differentiation between "class on paper" and "realized class" (1985, 1998). I argue, here, that tech workers constitute a contradictory and thus only partially realized class, shaped by contextual experiences of multiple crises and of tech workers' different socio-structural backgrounds and positions. This chapter also systemizes previously reconstructed national differences between tech workers in the US and Germany as well as professional differences between data scientists and UX designers in terms of class formation. Chapter 6 moves to examine how some of tech workers' social codes are in the process of being organizationally appropriated and co-opted, thus contributing to yet another spirit of capitalism. This chapter delineates three modes of appropriation—*corporate humanism*, the *play of differences*, and *compassionate design*—through which tech workers' post-entrepreneurial codes contribute to a moral rejuvenation of capitalism. Chapter 7 then zooms out further to explore the contradictory class identity of tech workers within wider societal contexts and the global class matrix. By mobilizing the concept of the "glitch," this chapter helps us to understand how tech workers' social codes can simultaneously bridge and deepen divisions within the contemporary middle class. This analysis extends to considering the possible making of a transnational class of tech workers. The book closes by weaving together the political and social theoretical implications of this study and mapping the critical pathways this research opens up.

2 THE RETURN OF SOCIAL CRITIQUE

On October 3, 2021, Frances Haugen, an American data scientist, revealed her identity as the whistleblower of "The Facebook Files," a series of reports detailing Big Tech's failure to prevent fake news and algorithmic discrimination as well as the negative mental health impacts of social media. Haugen's actions led to congressional hearings and public debates about splitting up major tech companies. This wasn't the only incident that shook Silicon Valley in 2021. Only a few months earlier, on January 4, 2021, a range of employees and contractors at Google's parent company Alphabet announced the formation of the Alphabet Workers Union. Internet tech industry employees had been thought of as non-unionizable—but by 2023, the Alphabet union reported more than 1,400 members and defined itself as continuing to strive "to protect Alphabet workers, our global society, and our world. We promote solidarity, democracy, and social and economic justice" (Alphabet Workers Union 2023).

Disruptions like these shouldn't be written off as isolated events: They tell us something deeper about how tech workers see the world. To understand the social making of these events and why its well-paid cadre of coders were stirring things up, we need to home in on tech workers' worldviews and decode these. Understanding worldviews as clusters of values about the state of the cosmos that one inhabits (Kalberg 2004, 140; Weber 1968, 450), I find the majority of my tech worker interviewees to have a strong normative focus on challenging social inequalities. If the neoliberal ideal of the entrepreneurial employee is mostly unconcerned with social critique—oriented instead toward market logics and artistic principles of creativity and self-actualization (Boltanski and Chiapello 2018)—these tech workers channel their normative capacities into (1) a critique of economic inequality, (2) a critique of insufficient diversity, and (3) a valorization of reflective data work. These moral codifications go hand in hand with symbolic boundary-makings based on distinguishing themselves from the entrepreneurial class as well as other professional groups.

Decoding the worldviews of tech workers reveals the considerable potential they hold to challenge tech leaders and steer innovation economies toward digital tools that better serve the public good. Yet the return to social critique evident in tech workers' subjectivities shows limitations. A close examination of their schemas of perception and recognition illuminates the ways in which their moral codifications often remain anchored in symbolic politics and individualistic paradigms that hinder widespread or large-scale collective organization and union-building. This tendency illustrates a broader pattern: a partial transformation of the entrepreneurial self that falls short of a genuine rupture and offers an explanation of why there have been relatively few organized protests by tech workers against recent layoffs as well as the close cooperations between tech leaders and the second Trump administration.

After mapping out the social codes that emerge from interviews with tech workers, the fourth section of this chapter turns to consider the extent to which the institutional discourse about tech workers reflects these codes. This reveals significant tensions between how tech workers see themselves and how institutions imagine—or want—them to think and behave.

While stark differences emerge, however, there are also notable parallels suggesting that institutions have borrowed and repurposed certain elements of tech workers' social codes for their own ends.

2.1 CRITIQUE OF ECONOMIC INEQUALITY

The first form of social critique performed by tech workers is the *critique of economic inequality*. This critique is based on viewing capitalism as a labor-reliant system whose fruits are distributed unfairly. This schema manifests, for instance, in the following passage from an interview with George, a senior UX designer from a US start-up:

> We are in probably one of the most interesting times. I don't think we've seen this probably since the '30s or '40s, where there is a massive push towards socialism. And that comes because Walmart doesn't pay, you know, even a decent wage. And, I read yesterday, the average CEO salary grew 1,000 percent over the last thirty years. The average for workers grew 16 percent. That disparity and income inequality is getting Jeff Bezos worth $180 billion [. . .] Bernie Sanders' example is that Walmart pays its employees enough that they have to seek out help from the government for like food stamps, things like that. And yet the Walmart family is the richest family in the world. So, where's the problem? That problem is big here in Silicon Valley. And when you take on Amazon, and you want to unionize the employees, they're open to it because they know, Jeff Bezos, he's making an obscene amount of money.

George presents elaborate historical knowledge about rising income inequality. Furthermore, he connects this with a critique of capitalism in general, and the Walmart family and Jeff Bezos, the founder of Amazon, in particular. His social critique carries a "class-based view of society" (Thompson 1966, 9). In George's perception of society, there are rich families and corporate bosses on one side and workers, unions, and socialist politicians on the other. He cultivates a dichotomous worldview. George tells me all this via Zoom from his house in a suburb of San Francisco, where our interview is sometimes interrupted by the family dog, a golden retriever. He strikes me as a settled person, at ease with his personal situation while simultaneously critical of society at a fundamental level.

It's important to acknowledge that not all of my interview partners expressed such elaborate and informed critiques of economy and society. However, a critique of economic inequality was frequently voiced in some form. Mike, a US data scientist I interviewed spoke of "the middle class being squeezed," and Bernardo, a data scientist who migrated to Germany from Brazil, even told me, "Personally, I think we live in a plutocracy for sure. I mean, we have a bunch of people that run society, dozens of families that control everything and that have very strict ties with the political power for sure." Halfway through my interviews, I began to realize that *conflictive worldviews* are typical for tech workers. For most, the social world appears not as a flat network but rather as a hierarchical system. This finding challenges public notions as well as some social scientific accounts of middle-class knowledge workers as class-blind or uncritical of the economy (Beck 1992). While neoliberal ideals have been diagnosed as dominant in the professional fields of finance, consultancy, and accounting (Lupu and Empson 2015; Neckel et al. 2018; Neely 2020; Schmidt-Wellenburg 2013; Spence and Carter 2014), tech workers present themselves as more critical of contemporary society.

At times, the return of the social critique among tech workers also leads to problematizations of the social situation of lower social classes. A number of my interviewees articulated concerns regarding the social situation of low-paid digital laborers such as gig workers. Max, a freelancer and UX designer for a German start-up, expressed empathy with the precarious situation of gig workers:

> They're the ones being used by the tech industry. They're a physical-world extension of the high numbers that you can get out of doing things digitally. [. . .] The gig economy is a dehumanization. I mean, they're literally being replaced by robots. Like, they're not considered humans anymore, like, it's super shit.

Max ties the highly precarious employment conditions of gig workers to the digitalization of the economy. Even though Max describes himself as a rather happy "digital nomad," for him, digital technologies can also be dehumanizing. Most tech workers who talked with me about gig workers expressed disapproval around their social situation. This shows that tech workers do not simply buy into a "technological solutionism," understood as an uncritical belief in the progressive nature of digital technologies, which is common among tech entrepreneurs (Morozov 2013; Nachtwey and Seidl 2020). This is not to say that they are not deeply invested in hopes of the progressive potential of technologies. Furthermore, they rarely reflect on the responsibilities of their profession in programming and designing the digital technologies that surveil and control gig workers. However, they are in touch with contemporary discourses that problematize the tech industry and display a notable awareness of harms associated with digital technologies. They cultivate a codex based around not only being politically and economically informed but also being both critical and caring.

Ideal Types

It must be highlighted, here, that the degree to which tech workers cultivate codes of social critique and concern varies substantially across individuals. Tech workers do not constitute a homogenous bloc, meaning it is important to elaborate on the different ideal types of tech workers before delving deeper into their social codes. The concept of ideal types operates within abstract and accentuated models that help us to analyze complex realities (Weber 1904). Three ideal types of tech workers can be differentiated: *the radical tech worker, the reformist tech worker*, and *the affirmative tech worker*. Notably, empirical individuals are never pure manifestations of a single type. Instead, they tend to lean more toward one type than another, though their orientations can shift dramatically depending on the situation.

At the heretic pole of the field of tech, roughly one quarter of my interviewees lean toward the radical type. The radical tech worker, who prefers working at start-ups, carries a deeply conflictive worldview. Individuals attributed to this type often speak the language of activism, including Rachel, the interviewee we met in the prologue, who explicitly supports protests against employers. Notably, a certain portion of this ideal type consists of tech workers from the Global South who managed to obtain a secure visa status. Radical tech workers are highly sensitive to modes of domination. Politically, they often perceive a power elite at play, like Bernardo when he talks about "living in a plutocracy." Without being revolutionaries or genuine anti-capitalists, individuals who resemble the social figure of the radical long for structural reconfigurations of economy and society. They are open and sympathetic to alternative political concepts, such as the "socialist vision" of Bernie Sanders, and they generally cannot imagine working for a so-called Big Tech company. Furthermore, they are the most likely to report engaging in collective action such as walk-outs or other acts of workplace solidarity.

However, this ideal type is not the most widespread. The majority of tech workers I met are best described as reformist. The reformist ideal type is no heretic but rather a heterodox actor; this type longs for incremental rather than radical change. Reformist tech workers accounted for approximately half of my interviewees. They are the most ambiguous type around issues of economy and society. On the one hand, they are skeptical of demands to split up Big Tech. Furthermore, they are often uninformed about protest movements and unionization efforts in tech. On the other hand, reformist tech workers do indeed perceive an array of problems with contemporary capitalism that require fixing. David, whom we encountered in the prologue, leans toward this type and shares with radical tech workers a sense that there are serious discrepancies between how the world is and how it should be. Reformists signal a moral character by frequently problematizing economic inequality and a lack of diversity. However, the ideal typical reformist tech worker is less fierce in doing so, and ultimately considers problematic issues to be fixable within the given socioeconomic principles. Reformist actors ultimately stand for incremental rather than structural change.

Last but not least, there is the affirmative tech worker. This type, located at the orthodox pole of the field of tech, works at both start-ups and established tech companies and is generally d'accord with the modus operandi of the tech industry and society. Affirmative tech workers accounted for roughly one quarter of my interviewees. Quite often, tech workers leaning toward this category aspire to one day become start-up founders themselves. While they are not entirely devoid of critical positionings, they do not carry a conflictive or hierarchical worldview and tend to distance themselves from the public scrutiny that has hit the tech industry since the Cambridge Analytica scandal in 2016. Quite often, they favor conservative political parties over liberal or left-leaning ones. Furthermore, affirmative tech workers mostly channel their normative capacities into self-orientations rather than moral other-orientations. The affirmative tech worker comes closest to the subjectivity associated with the neoliberal ideals of the entrepreneurial self or the *Homo oeconomicus*.

While it is important to acknowledge this heterogeneity in terms of the three ideal types, I also want to underscore the fact that the majority of my interview partners appeared to be genuinely critical of the state of the economy and society. Roughly three quarters of my sample leaned toward the reformist or radical ideal types, and their dissatisfaction translates into a more or less pronounced transfiguration of a holistically market-oriented selfhood.

Social Critique and Subjectivity

The departure from post-Fordist work culture is perhaps most explicit in my interviewees' general self-classifications as "workers" rather than as "entrepreneurs" or "professionals." Most of my interview partners either identified as "tech workers" without me asking or self-identified as such once asked. This matters because it indicates that tech workers are not fully taken in by a "Californian Ideology" that glorifies entrepreneurialism and individualism (Barbrook and Cameron 1995). Most of the white-collar folks I met would be mischaracterized if we painted them as "wannabe entrepreneurs."

But who exactly counts as a "tech worker" for my interviewees? One illustrative answer was provided by Sebastian, a French data scientist in his late twenties who works for a large fintech company in Berlin. Sebastian, who earns €85,000 a year and wears a plain black t-shirt when we meet on Zoom, confidently identifies as a tech worker. When I asked him what this means, he tells me:

> **Sebastian:** I work in tech. So I'm a tech worker.
>
> **Robert:** And who are, like—who are the groups who count as tech workers for you?
>
> **Sebastian:** Everybody working in, like, primarily a tech company. So, like, if you are a product manager in a tech company, you are a tech worker. So, I mean, it's basically all the people who are, like, super privileged, you know, who just work on tech products. So, like, product teams usually. Like, yeah, product teams, they have like you know, a product manager, a designer. So even if designer is, like, more like creative [. . .] they are still tech workers if they work in, like, tech companies [. . .]
>
> **Robert:** And what about like, social media managers? People like that?
>
> **Sebastian:** Yeah, I mean, I guess if you work in a tech company, like you're an employee of a tech company. But if you work in a museum, and you're a social media manager, I don't think you're a tech worker. [. . .]

Robert: And in your eyes, the people working at the call center at your company, would they be tech workers?

Sebastian: Well, actually, that's an interesting question that you ask because, like [company name] put them, so most of them are actually based in Berlin. But, like, they typically belong to like a different organization. [. . .] they don't have the same laptops, don't have the same policies, don't have the same everything. Because they typically have like, shorter contracts. So in that case, I think there are like boundaries.

Sebastian understands tech workers as those employees at tech companies who can be considered "privileged" and who "work on tech products." Two types of boundaries are present in this classification. First, Sebastian draws a distinction within the field of tech. He considers only those tech employees who are privileged in terms of work contracts and benefits to be tech workers. On the grounds of material and institutional differences, he does not see low-paid call center workers at a tech company as tech workers.[1] Secondly, Sebastian distinguishes privileged employees at tech companies from privileged employees at non-tech companies. Even though a social media manager at a museum may be privileged to a certain degree and may also conduct very similar work to a social media manager at a fintech company, only the latter is considered a tech worker. For Sebastian, the term "tech worker" is linked to the field of tech companies as well as to privileged positions within that field.[2] While not all tech workers are this explicit about the distinction, a number of my interview partners more or less echo this classification scheme. Levi, for instance, a Dutch UX designer from Berlin, told me he considers a tech worker to be "anybody that works in start-ups and environments that make the digital, that shape the digital environment. I guess. So, I wouldn't say that's someone that works with tech digital tools, like Google Docs, or whatever. But rather if the aim, the objective of your work, is to make those products." For Sebastian and Levi, you don't work in tech simply by using tech products. Following this principle of classification, even a world-class coder may not possess the entry code for the realm of tech workers if he works in an economic field that is considered genuinely unrelated to tech companies.

The classification of "tech workers" thus functions along both exclusionary and inclusionary axes. It operates on exclusionary principles in the sense that it distinguishes digital laborers from nondigital laborers. Furthermore, it also distinguishes privileged digital laborers from precarious digital laborers (this contrasts with how some activists and researchers deploy the term "tech worker"). At the same time, though, it's important to note that tech workers nevertheless classify themselves as "workers" rather than as "professionals" or "entrepreneurs." Tech workers cultivate a code of *workerism.* They draw symbolic boundaries vis-à-vis the actors who buy their labor power. Workerism is a surprising codification among actors who have been long considered unionizable (Tarnoff 2020). At the same time, it must also be acknowledged that not all social codes that configure the entrepreneurial self are entirely rejected by tech workers. Mobility and flexibility are still highly valued. Sebastian, for instance, told me he likes to change jobs "every three years or so" in order to challenge himself. Other tech workers I interviewed highlighted the importance of autonomy and self-fulfillment. The neoliberal schemas of flexibility, improvement, and expressive individualism have not evaporated but still find a place in the hearts and minds of tech workers. However, these issues were not a dominant source for the normative orientations of tech workers. The majority of tech workers I met are currently satisfied in terms of their capacity for flexibility or creativity. It is exploitation, rather than alienation, that concerns them primarily. This way of perceiving the world undergirds their self-classification as workers.

Another way the social critique manifests among tech workers is through *openness toward more government regulation.* The entrepreneurial self generally favored the market as a regulative principle and was skeptical of government regulation.[3] Tech workers, in contrast, are quite open to regulative and interventionist state policies. Attuned to public discourses that problematize the power of Big Tech, most of my interviewees renounce political beliefs around deregulation and laissez-faire markets (Brown, 2015). This is illustrated in the remarks of Xavier, a Peruvian data scientist with a German start-up:

> Big companies, yeah, for example, Amazon, I'm always, I'm not too happy about the big power that it has, right, like, which is, which we could consider in some places a monopoly. And I think it's really time that it's being looked at directly to see whether the company needs to be—and also Google, right—whether they need to be split, because they definitely have, like, a lot of power.

If we conceptualize neoliberalism in the economic sense, such that the task of the government is to secure competition in markets, then this utterance can actually be interpreted as a genuine neoliberal statement. Some of my interviewees favored government interventions specifically in order to secure a competitive market system. However, if we understand neoliberalism rather as a "conduct of conducts" (Foucault 2008) that has interpellated subjects to be holistically oriented toward the market and to be generally skeptical of market regulations (Mau 2015; Rose 1989), then this utterance illustrates a disillusionment and skepticism with the larger neoliberal configuration. Most tech workers I interviewed believe that governments should play a stronger regulative role. Furthermore, tech workers are also open to more redistributive policies. Beyond more regulation, higher taxes are considered favorable among tech workers who lean toward the radical and reformist type. Matteo, a data scientist from Germany, told me:

> It bothers me that workers have less and less wealth. Because working is not a big deal any longer. I mean, having a job is not like in the '90s. And I am worried if the state doesn't compensate with taxes, because otherwise, [those] who have money will increase their property and their profit, and then you increase the gap. [. . .] If the state doesn't rebalance with taxes, it is going to be a disaster.

Matteo, who is originally from Italy, can be considered as between the ideal types of the radical and reformist tech worker. He is concerned about the value of work and the liberalization of the economy. Even though he works at a big tech company in Berlin, he problematizes the distribution of economic capital and expresses a longing for more stability and equality being provided through the state. In my interview with Matteo, I also asked him whether he would recommend the profession of data science to his children. He replied:

> Yeah, I mean, I motivated my cousin. She did engineering, math engineering. And now she's doing a PhD in data science. So I'm glad for her. I think it's a good choice. But I mean, by the time my children would be 18, 19, who knows what will be the job, you know, and the market situation. And I guess it's harder for this generation now. To jump into university, not knowing the market when you would be done. It was hard for me, and now, really, it is even crazier with this market that is so unstable, unpredictable.

Matteo desires a more stable economic future for himself and his family. He uses the word "crazy" to distance himself from uncertain labor market futures. Like most of his peers, Matteo is worried about the state of the economy and where it is going. Reflecting on the economic field, he describes a state of anomie: a system in flux, volatile and unsettled. Matteo longs for state intervention and therefore cannot be understood as a subject who is completely captured by a neoliberal ideology. Even though tech workers belong to a rising fraction of the middle class, whose economic earnings clearly put them in its upper segment, they are worried about the market. This concerns not just the labor market but also the housing market. Many tech workers I met experience the housing market as a threat to the social fabric of society as well as to themselves, despite their high incomes. Marco, a tech worker and former sociology professor who now makes more than $200,000 a year, told me he would like to go back to academia but was hesitant because of the increasing high housing costs in the Bay Area. He described how "to live here, I have to have jobs like Amazon" [company name changed]. This statement illustrates how tech workers can become confined to certain career paths; the housing market and the mortgage system can create *lock-in effects*. Marco is worried that leaving his tech career could threaten his middle-class way of life. His concerns are not unfounded given that a family of four living in Silicon Valley in 2019 was estimated to require $131,600 to meet their basic needs (Chen 2022, 205). In Berlin, too, a number of tech workers told me that they fear that soon, even for them, it will be unaffordable to buy an apartment in a central location. Real estate prices have more than doubled there over the last ten years. Tech workers aspire to a middle-class lifestyle that generally includes property ownership. The soaring prices for houses and apartments must be considered as a central factor that accounts for the prevalence of social critique among this group.

Another social code that signals a transformation of the neoliberal figure of the entrepreneurial self among tech workers can be located in their quest for a *redifferentiation between the spheres of work and life*. Tech workers long for a life outside work and report relatively regular work hours, expressed in the following statement by Emma, a senior data scientist from a big tech platform company in the US:

> I work from nine to five. Then from five to seven, it's, like, family time. And then afterwards, I'll maybe, depending on if I need to or not, I'll work one or two hours in the evening. And I only do it if it's necessary. Like I am not someone who is going to be, okay, well, I'm not winning any awards for the hours that I'm working, I guess. I just get the work done that I need to do. [. . .] I mean, I have some colleagues that like, are up, especially like, some of the higher-up people or the like, older [company name] people, that work really crazy hours. But I don't feel like I need to do that to do my job. [. . .] I'm not going to compromise on the time with my family for work. [. . .] Also I have friends, like in consultancy, that work really crazy hours, you know, and I just don't know why they do it.

Emma is far from radical in her critiques of the market. She cultivates elements of the affirmative type when she tells me that she feels the critics of the tech industry dwell too much on its negative aspects. Nevertheless, Emma exhibits a tamed work ethic and seeks a redifferentiation between work and life based on cultivating a positive revaluation of being a "normal" employee. Her subjectivity moves across the delineated ideal types depending on the life sphere and the situations that come with it. In terms of work hours, she distances herself from older generations of tech workers as well as from friends in consulting, considering their hours to be "crazy." For Emma, who reports working forty-five hours a week, the sphere of work is a means rather than an end in itself.[4]

I detected such a strong commitment to the private life among most of my interviewees. Similar to Goldthorpe's classic study of the "affluent worker" (Goldthorpe et al. 1967), I find a high valorization of family life among my interviewees. However, it must also be noted that most tech workers combine the subjectivity of being "normal employees" who sell their labor at a high price with the cultivation of social critique. This code of conduct can also be located in the social code of *workplace collectivism*. Tech workers cherish assisting and informally helping each other out at the workplace. This includes sharing information, especially computer code—a practice that corresponds to values of the hacker community and finds an institutional manifestation in the open source movement and open source platforms (Marwick 2013, 33)—as well as salary information in a few cases.

Norris, a data scientist from the US, told me he was supported by a colleague when negotiating his pay with the employer:

> I was going to move from about $75,000 to $85,000 because I was moving into his role. And he told me what he made, about $105,000, which allowed me to demand that I be paid what he was paid. I firmly believe that pay transparency is big and important and huge. I think that it's great for us to be able to share that [information] with other people where possible, because, especially going back to equity and inclusion, I think it's, you know, it tends to be companies who benefit when we don't share salaries, and then white men tend to benefit the most, followed by white women, then, you know, men of color, women of color, and so on. And so I think that the more that we can share and be transparent about that, the better it is and the more we can help create a more equitable, you know, future for our colleagues in in the tech industry and beyond.

Social critique here manifests in this reported practice of helping each other out with information to secure a larger share of their companies' profits. This illustrates that social critique has the potential to manifest not only in individual responses, such as managing one's work–life boundaries, but also in collective action, which takes place when two or more actors join forces to pursue their interests together. Norris thinks and speaks of himself as part of a collective of employees with distinct interests. Like many other of my interviewees, he told me the financial crisis hit when he entered the labor market and ultimately led him to pursue a career in the promising field of tech. The financial crisis constitutes one important context that accounts for tech workers' disillusionment with neoliberal market ideals. In terms of ideal types within the tech worker grouping, Norris leans toward the radical type. In his worldview, there is a high degree of difference between the interests of tech workers and the interests of the firm. Interestingly, Norris deploys the concept of "sharing" and turns it against the tech companies. Furthermore, we should note that Norris also expresses concern for diversity—a codex that I will explore further in the next chapter. Nevertheless, it is important to acknowledge that reports of collective action are rather rare among tech workers. Furthermore, even those who do occasionally engage in collaborative efforts typically do not use established institutional channels or bodies to do so. Looking at Norris's career on his LinkedIn page, I noticed that he changes employers every few years. This may partly explain why Norris is not a member of a union even though he told me he could imagining joining one (he does not discuss the possibility of establishing one).

At this point, I will explore in more depth the linkage between social critique and politics. I hold that the social codes around being critical about the economy and society, cultivating workerism, redifferentiating between work and life, and engaging in workplace solidarity all correspond with a specific *political subjectivity* of tech workers. In Germany, the majority of my interview partners (whom I recruited through different channels) identified as "left-wing" or "green." Levi, for instance, told me, "I would say, I'm a left-leaning socialist. I end up voting for the Green Party a lot of the time. Not necessarily socialist parties, I think that usually goes [pause] yeah, but sort of a bit center-left." Levi is skeptical of radical political projects, instead preferring change within the given socioeconomic configuration. In the US, the majority of my interview partners identified as "liberal," "democrat," or "left-wing." Tara, a UX designer, ties her political orientation to her profession and aspiration to improve the world: "I used to think I was conservative, but I really think I am more democratic leaning. Just because, I don't know, to me, I choose to be a UX designer because I want to do good things for others. And I'm not seeing that from the conservative party here." Around a quarter of my interviewees identified with the categories "very left-wing," "socialist," or related political affiliations, demonstrating their longing for major political change. These interviewees more or less resemble the radical type. One of my interviewees who leans toward this type is Craig, a data scientist at a start-up in Boston, who expressed sympathy with candidates such as Bernie Sanders or Elizabeth Warren in past elections:

> **Craig:** I would say I'm liberal. Like, American liberal.
> **Robert:** American liberal would be Biden, but not Sanders?
> **Craig:** Yeah, that's a good distinction. Alright, so I feel like Biden is pretty like liberal moderate. I would say that I'm more liberal than Biden. Universal health care would be great, which is one of Sanders' positions. So, I guess I'm saying I'm more Sanders than Biden.

This politically left-wing or liberal identification, in combination with the other norms and logics cultivated by tech workers previously discussed, offers an account of the forms of subjectivity that have enabled tech worker activism. While tech was long considered an anti-union world, it is now precisely in Silicon Valley, the heart chamber of digital capitalism, that recent unionizations have taken place (Nedzhvetskaya and Tan 2024b; Tan et al. 2023; Tarnoff 2020). As well as at Alphabet,[5] tech workers have unionized at the tech company Kickstarter. In Germany, there have also been attempts to unionize tech workers, for instance at the fintech bank N26. Activist scholars have coined this development as the beginning of a "Tech Worker Movement" (Tarnoff 2020; Tarnoff and Weigel 2020). It has yet to be determined to what extent this movement will continue to thrive. While the majority of my interviewees said they could imagine joining a union, only two of my interviewees reported being directly involved in unionizing practices. While this insight does not mean that there is no collective action, since this can take place outside of unions, it does help us to better understand why we saw so few coordinated protests against the layoffs at Big Tech companies in late 2022 and early 2023. The tech workers I met are sympathetic to unions, yet they also state that it is difficult for them to imagine what a union would look like at their workplace. Furthermore, most of the tech work unions that have been founded thus far are not yet formally recognized under governmental law. As such, it remains to be seen how far the return of the social critique will go in transforming the mass of white-collar tech workers into a red-collar social class. Tech workers are a middle class fraction in the making.

At the same time, this analysis has revealed that neoliberal ideology has lost dominance over the ideal identity of white-collar middle-class workers. Tech workers' conflictive worldviews, inclusionary repertoires of identifying as "workers," revalorization of regular work hours, workplace collectivism, and openness toward more government regulation and unionization signal a renunciation of key aspects of the entrepreneurial self with its highly capitalist orientations. Despite the ambivalences and contradictions that remain, this identity signals potential for collective organization as well as new class alliances and is particularly relevant given that tech workers constitute one of the few rising middle-class fractions in the context of neoliberal economies.

The insights throughout *The Social Codes of Tech Workers* reveal possibilities for broader social change that could challenge the entrenched norms of the neoliberal order. In the following section, I will explore how tech workers trouble the ideal of the entrepreneurial self as well as the paradigm of technosolutionism through a social and normative codex that centers around a concern for diversity.

2.2 CRITIQUE OF INSUFFICIENT DIVERSITY

It is not only economic inequality that concerns tech workers. As well as rejuvenating the classic social critique, they are also engaged in *reprogramming* the social critique through their commitment to diversity. In particular, tech workers enact critiques of gender and racial discrimination. The importance they grant to diversity (which is, of course, related with, but not equivalent to, concerns about economic issues) marks an extension of the classic social critique. The normative codex of tech workers includes a critique of economic inequality and a critique of insufficient diversity. Traditionally, the social critique among middle-class workers was primarily in the interests of male and white subjects. It was rarely spun to broaden access to the labor market or tackle discrimination at the workplace. This led to a neglect of the interests of women, People of Color, and other marginalized social groups, who profited only indirectly—if at all—from the distributive effects of the social critique (Fraser 2009; Gregg 2012). The new social critique is more inclusive. It extends its reach through problematizations of the underrepresentation of women as well as People of Color in the tech industry. This concern around gender diversity is expressed in the following comments by Meiling, a Chinese designer at a big tech company in the US:

> I do notice I'm the only woman on my team. During a meeting, I'm the only one. Yeah. Yeah. So, I feel it's not enough women in the tech industry. Um, yeah. I don't know why. [. . .] There's just more guys at school for a computer science major. So there's more men workers. But for UX designers [. . .] I remember at school there's more girls than guys. But at the workspace there is, like, more guys than girls.

This statement demonstrates Meiling's awareness of the stark gender imbalance within the corporate offices of the tech industry. She reflects that most of her fellow design students at the university were female, but in the workplace, most are men. She problematizes this state of affairs while also seeming somewhat puzzled about it. A great number of my interview participants, clearly in touch with discourses in the public sphere, critiqued the lack of women in tech.

But tech workers do not think the *lack* of women in the workplace is the only problem. In my seventeen interviews with women, three of them reported having experienced or witnessed *discrimination* against women due to pregnancy (notably, I had not asked any questions about pregnancy issues). For instance, when I asked Britta, a German UX designer at a US start-up, whether her gender plays a role in her industry, she told me:

> It sure does play a role. So, there are way less women in tech, especially in automobile. And I have the feeling that regarding gender equality we are way ahead in Europe. For example, one colleague of mine was pregnant, and what happened was that she was no longer included in projects and also not considered for promotions. She was discriminated [against]. I don't think something like that would happen at German companies. Or it would be illegal. I think regarding diversity, a lot more can happen. There should be more women in tech companies.

Britta reports discrimination against a colleague of hers. She connects this with the argument that in Germany, discriminatory behavior toward pregnant women would not happen, or at least is less likely to happen (it is important to note, however, that one female German tech start-up employee I interviewed reported pregnancy-related discrimination). Emma, the one US tech worker who reported a direct experience of discrimination due to taking maternity leave, told me, "I guess the one time that I felt discriminated against is when I was on maternity leave with my second child, and I was up for promotion. But then my promotion got deferred until after I got back from maternity leave." To gain knowledge around how widespread such practices are across different types of tech companies and national sites, quantitative research is needed. However, it appears that the tech industry, which has tried hard to promote an image of being open toward women, is still a discriminatory habitat, particularly for women who want to have both a career and a family.[6]

The outward image of the tech industry as an inclusionary world of work is further challenged by the phenomenon of "techno-masculinity" (Poster 2013). While most of my male interviewees did not present themselves as "tech bros" (Carrigan 2024; Lobo 2014), it is important to highlight that *latent forms of techno-masculinity persist*. These forms manifest especially through a gendered hierarchy between professions. I found that the profession of data science, where men are particularly overrepresented, tends to codify itself as more "technical" in comparison to the profession of UX design, where women are less underrepresented, but which is framed as more "creative" within the segment of tech workers. For instance, Mike, a data scientist from the US, told me "I'm a bit more technical" when talking about other professions, including UX designers. Such coding practices contribute to the fact that seemingly more creative tech workers, such as UX designers, earn significantly lower average salaries than those in more ostensibly technical roles, such as data science.

This coding of a symbolic hierarchy within the field of tech work reproduces gender inequalities through the differently valorized attributes of "more technical" and "more creative." Interestingly, the mapping of gender onto the binary code of technical/creative is a historically highly contingent phenomenon in the field of tech work. In a seminal study of the history of computer work, Hicks (2017) shows how from the onset of computer work in the 1940s until the 1960s, computer programming was viewed as women's work. In this epoch, "technical work" in offices was coded as low-status and female, while "creative work" in office settings was coded as high-status and male. However, once computers gained more prominence in the 1960s and governments spun programs to create a male technocratic elite, the valorization scheme was reversed (Hicks 2017, 14, 149). Hicks's historical perspective complicates our knowledge of the meanings attached to computer programming and reveals how *masculine dominance in "tech" is secured through shifting valorizations of technical and creative work*.

The contemporary field of tech work is thus far from egalitarian. My work points to persistent gender inequalities that are generated through overt as well as more subtle or indirect forms of discrimination (see also Carrigan 2024; Neely et al. 2023; Sengupta and Tacheva 2022; Wu 2020). At the same time, however, I want to highlight the virulent critiques of insufficient diversity that are articulated in this context. There is an increasing awareness and critique of the unequal gender makeup of the industry. While discrimination in tech is far from eradicated, it is increasingly problematized by members of the middle-class workforce of tech workers (especially, but not only, by those who have experienced discrimination).

Importantly, the critique of insufficient diversity not only addresses the social phenomenon of gender. A *broad understanding of diversity* can be found among most tech workers. They argue that broader inclusion of marginalized groups is necessary. A very illustrative example of this argument was given by Charly, a UX designer from the US who believes that women and queer people are better represented than People of Color. They (Charly identifies as non-binary) problematize the lack of People of Color in tech:

> And regarding diversity, so I haven't directly worked with more than one person of color, in my experience so far, which is really awful. And it's, I think, messed up, not right. And what I intend to do about it is once I'm at a place of, like, having hiring power I would like to influence what I can directly, like, within my company, or companies, where I can. And it's definitely not accessible to all people, I think that it seems to be a lot more inclusive of people in the queer community. And I know that that has been an issue, especially, so I know that women in tech has been a big issue. And I think that like after, I don't know, five or 10 years, it's made pretty big strides. And it seems like there are a lot more women in tech, which is wonderful. [. . .] But I think the next big steps that needs to be taken are being inclusive of People of Color and adding more People of Color to teams—and having those viewpoints especially in UX design, because when we're designing, we're not just designing for white people.

Charly demonstrates not only discursive concern for marginalized social groups but also the future intention to transform discursive problematizations into practice. They tie their career plans to the self-promise of influencing hiring decisions in the interest of more diversity. Furthermore, they link this normative problematization with an economic observation: Charly argues that more diversity at tech companies could lead to the design of more successful products. We can flesh out this stance to argue that Charly considers diverse project teams and successful tech products to form an elective affinity. They consider different experiences—which they primarily connect to different racial, ethnic, gender, and sexual identities (they do not mention class or age)—to be beneficial in designing products for user groups beyond white people. Charly constructs difference as a form of cultural capital for the rendering of digital technologies. For them, it is not creativity but *a play of differences* that is the foundational cultural resource of the digital economy, where difference can be interpreted as the necessary condition for developing creativity. This schema of perception and recognition can be tentatively considered a building block of a yet another "spirit of capitalism" (Boltanski and Chiapello 2018; Weber 2005).

From another angle, the issue of insufficient diversity also surfaces in *problematizations of bias in digital technologies*. Many tech workers are concerned with the possibility of bias (re)production through their work products. In particular, data scientists (but also some UX designers) worry that algorithms may (re)produce the discrimination of disadvantaged groups. Michael, an Indian data scientist raised in the US, told me, "I have problems with the, sort of, the application of AI and machine learning right now by large Silicon Valley players that are, you know, happily building their personal biases into algorithms with no end in sight." Here, Michael exhibits awareness of what science and technology studies scholars refer to as *inscription*: the social encoding of beliefs and values into technologies. Michael draws social boundaries vis-à-vis tech actors he deems unreflective and displays skepticism around the progressive potential of digital technologies. Notably, though, when I asked whether we should, then, understand data and technology not to be vehicles for change, he disagreed. For Michael, data and technology are a problem as well as a (potential) solution: "No, I think you can use data. But if you're going to use it to promote more equitable hiring, you have to be more nuanced and more powerful in your use of

AI than just trying to find correlations." As I will elaborate on in chapter 5, the concern for bias is more pronounced among tech workers in the US than among tech workers in Germany. While also worried about data bias, the latter are more concerned with issues around data privacy.

In either case, tech workers should be described as *critical solutionists* in the sense that they believe in the power of technology while remaining critical of possible unintentional consequences that may result from the ways it is rendered. Especially the issue of bias is widely problematized (see Brown et al. 2024 on different understandings of bias among tech workers). Of course, taking issue with digitalization processes does not apply to all of my interviewees. Gopal, for instance, a UX designer from the US, did not express any concern about negative impacts of digital technologies when telling me:

> When I was ten, I was riding bikes. So, within three decades, we've come this far that a ten year old is able to pick up a camera, record a video and post it on YouTube. So another ten years from now, or even 20 years, it's gonna be a totally different thing that ten year olds will be able to do.

Gopal, like other more affirmative tech workers, shows that technosolutionist thinking still runs through the tech industry's ranks. Nevertheless, I found that most of my interviewees, professionals who navigate the complex relations between human and nonhuman actors, not only recognize their class position but also critically reflect on their inscription power. They demonstrate a relatively *reflective technosocial relation to the digital means of production.* I will expand on tech workers' relationship with technosolutionism in the next section. At this point, it is important to recognize the emancipatory potential opened up by these shifts in worldview. Through their critical assessments of the social making of digital technologies, tech workers signal the capacity to drive change from within the tech industry as they aspire to develop technologies that better contribute to the greater social good. Furthermore, their expertise carries special weight, meaning they could genuinely challenge entrepreneurs who often deflect critical debates about the effects of contemporary tech by questioning critics' technical knowledge.

That said, it is important to highlight the ways in which this code of displaying reflexivity about inscription can serve as a *strategy without strategists* (Foucault 1978, 132). Given the public pressure on the tech industry, in part due to its reproduction of biases, I hold that the display of moral values and political sensitivity allows those in the professions of data science and UX design to appear to be responsible subjects. Tech workers cultivate an ethos of responsibility. This self-presentation, in turn, allows them to be seen as suitable for the increasingly politicized task area of rendering digital technologies. Tech workers arguably hold an *interest in the moral.* While in Germany, concerns around privacy issues are more pronounced than concerns around diversity, in both national fields, we can make out a new moral relationship toward work. This can be interpreted as an implicit distinction claimed by tech workers in regards to competing occupational groups. Furthermore, the concern for diversity likely also distinguishes tech workers from other upper-middle-class professionals, such as in the field of finance (Neckel et al. 2018; Neely 2020), where no comparable problematizations appear to be taking place. Tech workers' display of ethical codes thus amounts to a form of symbolic capital in social space and in the system of professions in particular (Abbott 1988).

Aside from serving as a strategy without strategists, this concern with diversity also holds the potential for further commodification. More than one of my interviewees openly discussed how there is not only a moral but also an economic motivation behind bringing together people from different backgrounds to collaborate on innovation projects. As I will elaborate on in chapter 7, tech workers' concerns around diversity can be instrumentalized by tech companies as new commodification practices as well as enabling their (re)branding as humanist corporations (see also Irani 2019, 14). Another unintentional consequence may be that the concern for diversity could crystallize into a new cultural-political barrier, through which middle class subjects shut out people from lower classes, who may not be as aware of the various social codes (for instance in language), in order to appear ethical and moral. During the interviews, I often sensed that my interlocutors closely observed my reaction to certain statements to figure out if I shared their liberal political orientations—whether or not I belonged in their camp. The possible prevalence of such cultural-political barriers must be taken seriously in light of the low proportion of tech workers (across racial groups) with working- or lower-class backgrounds in the field.

However, while I want to take the distinctive and strategic aspects of the social critique seriously—as well as the unintended consequences it may have—I hold that we should not neglect the *space of possibilities for emancipation*. Like Irani (2019), I find that tech workers generally embody an honest longing for diversity, political purpose, and world improvement. Their hopes and dreams run deeper than notions of "solutionism" (Morozov 2013; Nachtwey and Seidl 2020) on the one hand, or "virtue signaling", on the other, suggest. The moral schemas of tech workers not only function as resources for boundary-makings but also point to issues that my interviewees sincerely care about. Furthermore, the concern for diversity has also, at times, translated into practice. Tech workers have conducted walkouts, and, not too long ago, fierce protests among tech workers erupted when Alphabet fired a Black ethics expert who said Google suppressed her research on bias (Wong 2020).

While the entrepreneurial self is heavily engaged in drawing boundaries vis-à-vis actors perceived as lacking the will to become autonomous or creative, tech workers are more concerned with drawing boundaries vis-à-vis actors who are deemed to lack solidarity or reflexivity. This boundary-making is undergirded by a new social code when compared with the codes of the entrepreneurial self, whose value systems oscillated around self-actualization and career success. Certainly, this is not to say that all tech workers cultivate a deep-running moral concern for diversity. As demonstrated, latent forms of discrimination that manifest in "techno-masculinities" (Poster 2013) still persist just as a number of latent strategies that constitute the flip side to the increasing relevance of other-oriented values. Nevertheless, it's important to recognize the transforming elements of tech workers' general codex. The subjectivities of contemporary tech workers signal a space of possibilities for both a reproduction of relations of domination as well as an emancipation from the various forms of social inequalities that went hand in hand with the neoliberal ideal of the entrepreneurial self.

2.3 A NEW RULE OF DATA AND CODE

A third critical worldview that structures the subjectivity of tech workers centers on a valorization of reflective expert data work. Tech workers deeply distrust decision-making processes that seemingly lack reflective data work. They want to see society run on data-based expertise, not gut feeling. While it is important to note that my interviewees do not cultivate a naïve techno-solutionism, they envision *a new rule of data and code*. For tech workers, the question of whether digital technologies will enfold positive societal effects is not determined by ideas of an autonomous life of the technology itself. They do not perceive all technology as a hammer and social issues as nails. Instead, tech workers highlight the labor as well as the reflective capacities that are necessary to propel "good" technological developments. Sophie, a data scientist from the US who works in the domain of health care, elaborates on this:

> I think it's really important to understand the data, to understand the process that generates the data [. . .] I have to make sure that I'm using the best quality data and account for any problems with the data. And that's especially important in politics and healthcare because the data that is of lesser quality is usually of people of lower incomes and people who are less well represented in politics. And so you know, if you are not accounting for data quality problems, then you're probably introducing or continuing the inequities that are in politics and health care.

This extract from an interview with Sophie underscores the importance that tech workers attribute to a reflexive rendering of data. As delineated before, tech workers are largely aware that data is not objective and may therefore (re)produce problematic societal effects. They acknowledge that developers can encode their biases into digital technologies. At the same time, though, Sophie also reveals a normative horizon in which reflective labor in combination with "the best quality data" available can produce positive societal effects. Such a horizon is not only present among data scientists. UX designers similarly see progressive potential in digital technologies if their rendering is conducted via reflective data work. Jordan from Berlin told me, "The whole point of view of UX is you can't really go on a hunch without considering the data surrounding it." Furthermore, David, the Irish UX designer we met at the outset of this book, told me, "So you never know for certain, but you speak to people and rely on research to get an approximate guess of what will work and you use that guess to form a hypothesis and you use this to test and [. . .] to validate your assumptions." In the case of David, we can make out a valorization of the small-scale research practices that are necessary for "good" digital technologies to emerge.

For tech workers, tech should be built through disciplined labor grounded in morally based reflective capacities. They distance themselves from what they consider to be unrealistic expectations of "artificial intelligence" or "machine learning." The tech workers I interviewed enjoy elaborating on the practical work and close-knit supervision that is necessary for something like "unsupervised machine learning" to run. They engage in de-fetishizations of many digital phenomena. This codification practice, interestingly, goes hand in hand with boundary making vis-à-vis tech leaders. As David elaborates, he understands "good tech products" to require research and data rather than leaders with beliefs:

> So, you may have somebody with a very strong voice at a leadership level, who believes something should be done, and a new product will be created, but can't articulate why, but he or she just believes in it. And I think such belief is dangerous. You know, we should be careful not to rely on our beliefs. We have research, we have kind of lots of different ways of ensuring that it's not centered around somebody's idea but centered around the actual need of the person that will use whatever is designed at the end.

David devalorizes decision-making processes that rely on "belief" rather than "data." Like other interviewees, he is skeptical of leadership that is not data-driven. The valorization of reflective data work fuels the making of boundaries vis-à-vis the top. Tech workers position themselves against a naïve technosolutionism as well as against a mode of "charismatic authority" (Weber 1968, 405). These are no apostles blindly following the gospel of tech entrepreneurs.

For some of my interviewees, data analytical capabilities even take center stage within their narratives of why they entered the field of tech. An illustrative example is Rafael, who left the social enterprise sector in Brazil to work in the US tech industry as a data scientist. When I asked him at the beginning of our interview, "How did you get working in tech?" he told me:

> At the beginning of my professional life, I worked for a social enterprise in Brazil. And what we did there was train people in disadvantaged communities like, you know, rural areas, Indigenous tribes, to create small businesses and stuff like that. [. . .] And I thought we were doing effective and impactful work especially because people kept inviting us to come back. But I realized that I actually did not know whether we had an impact or not. There was no way of measuring it, or at least we did not know how to. And I had no idea whether we're making a difference or not [. . .] So that kind of planted a seed of questions around like research and data for me, you know, like, how could I actually know whether what I'm doing is having results.

In line with the promises of "effective altruism" (Gabriel 2017), Rafael differentiates between well-meant and well-executed moral action. He draws a symbolic boundary, distancing himself from social enterprises, which he deems to lack an evidence-based approach. For him, it is crucial to gain data-based reassurance about his other-oriented work. The "coldly quantitative data," which Boltanski and Chiapello (2018, 71) had observed as an enemy to the artistic, post-Fordist spirit of capitalism, now serves as a vehicle to reach salvation. Rafael's case demonstrates that even though tech workers' moral worldviews may be linked to a growing concern with social issues across contemporary society, there is a specific, data-oriented return of social critique particular to this group.

The manifold valorizations of reflective expert data work not only generate boundaries vis-à-vis tech leaders but also spur boundary-makings vis-à-vis actors outside of the tech industry. Among many of my interviewees, there is a more or less latent distrust in activists, experts, institutions, disciplines, and governments that are considered to be lacking the data resources and data work skills to adequately fulfill the tasks of their jurisdiction (see also Burrell and Fourcade 2021). Such a logic was clearly present in my exchange with Thomas, a data scientist at a big tech company who holds a PhD in sociology from an Ivy League university. Thomas tells me he is happy with his transition from academia to tech. He considers the data resources of tech to allow for more interesting sociological work than could be done by sociologists within academia:

> **Thomas:** I try to convince people that there are interesting careers and interesting problem spaces for sociologists [in tech]. You know, if you do survey design, there's a lot of really great stuff on survey design that we're doing. [. . .] I mean you don't need to have a survey to ask people every five years what kind of music genre they're listening to. Here you can just like look at our logs data, and you actually see where they're listening at and what they're listening to. And you can actually, like, have really interesting analyses based on these things and actually help improve, like, you know recommendations [. . .]
>
> **Robert:** So, you're saying that one can do better social scientific work, at the moment, at big tech companies than at universities?
>
> **Thomas:** I would say, yes, in the sense that you have access to much better data. And, yes, in the sense that people have an interest in supporting data science work and, you know, basically like empirical research that is high quality and that is replicable. And that offers, like, kind of real data insights more than just, you know, cute stories [. . .] The main limitation, of course, is that we don't publish, or we don't publish very often. And whenever we want to publish, it's, you know, it's a process.

While Thomas appears somewhat dissatisfied with the limited possibilities to publish research, he considers his transition from academia to tech as one that has allowed him to harness the best data available. In his view, the academic discipline of sociology lacks the possibilities of data analysis of big tech companies. Interestingly, Thomas does not seem at odds with the fact that the more encompassing datasets of most tech companies are not publicly accessible. While he problematizes the lack and quality of data that sociologists outside of tech operate with, he does not consider, let alone call for, the sharing of tech companies' superior data. For him and many other tech workers, the honest will to contribute to a better world stops short of questioning the ownership of digitally captured human experiences. The privatization and public inaccessibility of increasing amounts of data is generally taken for granted (see also Thaa 2020).

This doxa echoes a larger paradox of the field of tech, which can be located in the one-way street of transparency and sharing. While tech companies regularly urge consumers to be transparent about their lives, they generally cultivate great secrecy about the inner processes of their firms. The rendered digital products and services that are often packaged in the form of gifts (Nachtwey and Schaupp 2022) are exchanged not only at the price of data-extraction (Zuboff 2019) but also at the price of technical blackboxing. The new rule of data and code that structures tech workers' worldviews breaks with a naïve technosolutionism, but it does not break with undemocratic forms of generating digital technologies.

2.4 SOCIAL CRITIQUE AND INSTITUTIONAL DISCOURSE

As delineated, the worldviews manifest in the self-understandings of tech workers mark a deviation from the ideal of the entrepreneurial self. In fifty-two explorative interviews, I found that tech workers are generally not so much concerned with alienation as with inequality and bias. Again, this is not to say that the entrepreneurial subject is dead. Tech workers still strive, to different degrees, for (further) flexibility, autonomy, and self-actualization. However, the interviews reveal a more dominant system of dispositions that manifest in a *critique of economic relations and a concern for diversity*. Furthermore, I have shown how they strive to establish *a new rule of the code* through reflective data work. In the following, I address the question of whether these self-understandings also play a role in the discourse about tech workers. Do such moral worldviews surface in study programs and job ads? Or is there a different normative dimension in the discursive interpellation of tech workers in the academic and economic fields?

Regarding the discourse in the economic field, my study finds pronounced differences but also occasional correspondences—with these points of connection opening the door for institutions to co-opt tech workers' social codes. In terms of differences, we can identify a widespread absence of the social critique of economic relations. Tech firms do not interpellate tech workers as critics of the current state of the economy or economic life in general. While it might not come as a surprise that tech firms do not engage in a critique of their own power or employment conditions, the dissonance runs deeper. As outlined, the return of the social critique in the self-understandings of tech workers is flanked by a positive revaluation of the figure of the employee with regular working hours and a differentiation between work and life. Yet, in the job ads, tech workers are primarily framed as entrepreneurial selves. While self-activation, autonomy, and creativity remain themes within tech workers' self-understandings, these cultures of subjectivation take up a much more prominent position in their discursive interpellations. For instance, under a job ad header, "So who are you?" the tech start-up Ongo seeks a UX designer who makes no distinction between work and life: "You bleed beautiful graphics, dream in UX flows, and eat behavioral psychology and HCI for breakfast" (UX, US, Ongo, start-up). Such heightened entrepreneurial interpellations of tech workers are most notable at start-ups. But established tech firms also interpellate their white-collar staff much more as entrepreneurs than as employees. Take an Amazon recruitment ad: "As a Senior UX Designer on this team, you will blaze a trail while having fun and creating customer value. This role requires ownership, autonomy, and an ability to deliver results. You enjoy working efficiently to build the right things with limited guidance" (Amazon, US, UX). This job ad clearly interpellates tech workers as entrepreneurial selves rather than employees. The imperative to deliver results without much guidance implies that work is not so much about regular hours or organizational embeddedness but rather about individual success. Moreover, the term "ownership" points to the fact that Amazon desires work subjects who do not have the mentality of an employee. Such talk related to "ownership" was not present, however, among the tech workers I met (including my one interviewee at Amazon). Nevertheless, the term "ownership" surfaces in a number of job ads.

The neoliberal figure of the entrepreneurial self is thus very much alive in the economic discourse. In some job ads—for both Big Tech companies and start-ups—the term even surfaces directly: "To be successful in this role, you must be driven, self-directed, entrepreneurial and focused on delivering the right results" (Microsoft, US, DS). This repertoire is present across the US and Germany: "The team cultivates an entrepreneurial spirit by encouraging risk-taking, transparently sharing successes, failures, and learnings, practicing continuous improvement" (SAP, GER, UX). The longing to enjoy the security and stability of the organizational self is rarely taken up by the discourse. While the job ads do invoke themes of community and teamwork, they do not thematize the working conditions of tech workers. Aside from salary ranges and lists of perks (such as fresh fruit or free massages), there is no clear information on work hours, contract length, job security, or career prospects. Given that ideal types are accentuated models, there will always be a degree of deviation when we turn to empirical reality. Yet, my interviewees don't simply diverge from the entrepreneurial self archetype. Rather, there is a fundamental dissonance between tech workers' self-understanding and their discursive interpellation within the economic field.

In the academic field, this dissonance is also present though it is less pronounced. Certainly, universities thematize job security, in a sense, by highlighting the golden career prospects of ongoing tech workers (US universities, in particular, are eager to justify high tuition fees via the promise of a safe ticket into the upper middle class). However, universities also lean into the semantic of entrepreneurialism. Jefferson University, for instance, brands their study program in the following way: "To compete in today's growing market, professionals must have the tools necessary to approach each task effectively, including practical interactive knowledge, technological skills, development, production and post-production knowledge, visual communication skills, and information literacy" (Jefferson University, US, UX). Here, Jefferson University mobilizes the fear of labor competition to sell its academic product. While the language of universities is not as heavily marketized as that of tech companies, study programs show a clear market orientation. With regard to the curricula of tech worker study programs, this market orientation also manifests in the fact that a number of universities advertise their study programs by offering "industry connections" (in the US, connections within the tech industry are primarily advertised, while in Germany, universities more often advertise their ties with manufacturing industries).

Hence, even though the subjectivation of tech workers as entrepreneurial selves is more pronounced in the economic field, the academic field shows surprisingly neoliberal interpellation patterns. Given that academia traditionally presents itself as pursing different economies of worth, there is a strong bend, here, toward market logics. On the basis of these study programs, it can be stated that the academic field of tech does not propel a meaningful reorientation away from the ideal of entrepreneurialism. While most tech workers are cultivating an emergent post-entrepreneurial subjectivity, study programs still construct their graduates very much in sync with neoliberal market logics. This prevailing of "academic capitalism" (Slaughter and Rhoades 1997) points to an ongoing neoliberal intrusion of social codes of the economic field into the now less autonomous academic field.

Let us now turn to the concern for diversity, which was widely present among the tech workers I interviewed. Unlike in the case of concern for economic inequality, there is a correspondence between the discourse and the self-codification as regards caring about diversity. In the job ads and study programs, tech firms and universities echo diversity-focused aspects of the social critique in an apparent effort to present their corporate culture as morally aligned with their potential recruits. Diversity is explicitly addressed in the form of *diversity statements* within the analyzed job ads. Tech firms underscore, at the end of job ads, that they specifically welcome applications from women, People of Color, and other marginalized social groups. This practice may change in the light of Donald Trump's reelection and the responses of tech leadership to his critique of DEI programs. At least in the early 2020s, though, a valorization of diversity played a central role in the industry's portrayal of its corporate culture. In a different and perhaps more notable way, I also found that the theme of diversity surfaces through the semantic of *needs* in job ads. Across the US and Germany, both start-ups and established tech companies interpellate tech workers as *discoverers of granular needs*. The nutrition company Hundred, for instance, advertises for data scientists who want to join their "health journey" by discovering the individual needs of users:

> [We are] a tech company in the online nutrition industry aiming to simplify the customer health journey by providing users with tailored monthly nutrition packs [. . .] Our data science team is in a special position to tackle problems

> across digital product (web, recommendations), operations research (inventory), and general behavioral prediction (probabilistic models, time series analysis). The data around our e-commerce products will interact richly with granular measures of users' health and wellness, a domain where machine learning is only recently starting to gain traction. (Hundred 2020, GER)

The mantra is that the labor of tech workers can do good by *discovering the various needs* of individuals/customers. On this model, data is understood as enabling an economy that strips the production system of standardization. Data scientists are called upon to explore the needs of an individualistic society via data work; UX designers are interpellated as discoverers of needs through more qualitative research. They are called upon, for example, to "create a highly differentiated and personalized experience for consumers that revolutionizes the way people shop" (The YES, UX, US). This construction of tech workers connects to a "solutionism" (Morozov 2013) and "eschatology" (Geiger 2019, 171). But beyond this, the semantic of discovering individual needs connects to diversity in the sense of a recognition (and appreciation) of difference. Notably, though, difference through the semantic of needs is tied to diverse *individuals*, while the semantic of antidiscrimination is often tied to respecting diverse *groups*. Nevertheless, I interpret this discursive interpellation as standing in harmony with the general appreciation of diversity by tech workers. In line with their moral compass, the discourse constructs personalization and customization as professional ethical convictions.[7] Against this backdrop, tech workers are figured as professionals able to align and harmonize ethical concerns over diversity with business concerns for profit by means of the personalization of economic production. Here, we can see the first contours of yet another spirit of capitalism emerge through the institutional co-option of tech workers' cherished values of diversity.

In the academic fields, the concern for diversity also plays a major role. Here, the mantra of discovering granular needs is similarly widespread. In addition, the academic field thematizes the role of possible bias. In the US, in particular, the subjectivation of tech workers consists of an imperative to consider any potentially discriminatory impacts of data work. In line with tech workers' self-understandings, German universities also thematize bias but place more emphasis on data privacy, while universities in the US tend toward problematizing bias. US universities are much more concerned with practices of *discrimination*: "Pressing issues include: the ways in which using data can subtly exacerbate existing systemic prejudices, such as through implicit algorithmic bias" (Georgetown University 2019). Especially elite universities within the academic field in the US position tech workers as reflective work subjects who will consider the impact of data and design work on marginalized social groups. In the case of Germany, study programs also link tech work to issues of diversity. However, in German universities, the main ethical linkage made around tech work relates to the dimension of *data privacy*. Universities present their study programs as preparing tech workers to be reflective around issues of individual data rights. At higher-ranked universities, there are specific mandatory courses that teach professional conduct around data: "Students are introduced to the technical, legal, and ethical issues of data security, especially when dealing with personal data or when planning experiments in Data Science" (LMU Munich 2020). While reflections on privacy issues and discriminatory practices can certainly overlap, they are not equivalent. The German academic field is grounded in a traditional liberal ethic, while the US academic field establishes an ethic for data science that is closer to what is often referred to as "identity politics" (Fraser 2009).

Finally, the academic fields present a rather divided echo when we turn to the new rule of the code that tech workers presented in interviews. This time though there is less of a national divide than a hierarchical one. While almost all my interviewees signaled appreciation of reflective data work to leave a naïve technosolutonism behind, it is must be noted that, especially in the US but also in Germany, primarily elite universities provide courses related to reflective data work. Harvard University, for instance, advertises its master's program in data science through the mandatory course on "Critical Thinking and Data Science." At medium- and low-ranked universities, courses in subjects such as "data ethics" or "ethical design" are far rarer. Elite universities appear to be in closer touch with current developments of the economy and demands by professionals, likely leading to a reproduction of hierarchical relations within the academic fields through a moral shift in educational offers. With regard to the economic fields, we can spot the trend that the moral ethos of tech workers is channeled into a new form of reflective technosolutionism. While some tech companies still mobilize old-school technosolutionist narratives (where digital technology appear as a hammer and social issues as nails), other firms have taken up a language that semantically links technosolutionism with the semantics of "indiscriminately" and "reflection," as this data science job ad from a start-up in Berlin indicates: "At Zattoo we already leverage a lot of our data (we process 6TB daily!) and still want to do more. We want to build day to day business decisions more on hard facts and less on gut feeling. [. . .] Zattoo enables our users to view indiscriminately and we reflect that in our team too." We can detect that tech workers' worldviews around a new rule of the code are recoded to a certain degree in the discourses.

Based on the analysis of discourse, it becomes clear that tech workers' self-understandings as critics of economic inequality, insufficient diversity, and unreflective data work find a divided echo across the academic and economic fields. On the one hand, there are important differences and dissonances. While tech workers strive to establish symbolic boundaries that distinguish them from a holistically market-oriented self, the economic and academic fields have different ideas for and about them. Tech firms (still) openly dream of employees who act as self-motivated and passionate entrepreneurs, while academic institutions mobilize the fear of labor competition to sell their programs. At the same time, the discourse analysis revealed an array of corresponding codifications across tech workers' self-understandings and the discourse, some of which are pathways for institutional appropriations and possibly even another spirit of capitalism. In particular, tech workers' concern for diversity and their belief in the power of reflexive data work seemed to have served as strategic entry point for aligning their return to social critique with capitalism's structural need for growth and legitimacy.

3 HYBRID PROFESSIONAL ROLES

It is possible to identify at least three popular tropes about tech workers in the public imagination. On the one hand, there is a concept of tech workers as "soldiers." According to this trope, tech workers are intermediaries who simply execute the commands of tech entrepreneurs and CEOs. They are envisioned as a subordinate mass that moves fast and breaks things as directed, strictly obeying the entrepreneurial and managerial class with a view to riding a start-up "unicorn" to a lucrative exit or otherwise ascending the corporate ladder. This framing portrays tech workers as a homogenous army lacking internal diversity or a voice of their own. Such a public portrait surfaces in a range of popular depictions of the tech industry—for instance, in the blockbuster film *The Social Network*, which casts Facebook's early tech workers as the passive members of Mark Zuckerberg's entourage. A second common trope centers on a depiction of tech workers as "nerds." This image is clearly present in HBO's show *Silicon Valley*, where tech workers come across as idiosyncratic, awkward figures who have serious difficulty connecting to people outside their own thought community. The show frames them as actors with high cultural capital in terms of expert and technological knowledge but low cultural capital in terms of communicative and empathic skills. Finally, a third public trope centers on the depiction of tech work as a meritocratic business. Both *The Social Network* and *Silicon Valley*, as well as many other media depictions (e.g., Davenport and Patil 2012), imagine the tech industry as a habitat that allows for social mobility and professional success based on merit. These popular narratives rarely thematize entry barriers or discriminatory cultures, naturalizing the lack of representation of individuals from marginalized communities. These stories construct tech workers as subjects who have secured their legitimate place within the (upper) middle class through hard work and talent.

These public portrayals of tech workers are misleading. The coders and designers of our virtual spaces are not a homogenous group of corporate soldiers, nor are they stereotypical nerds, and they are certainly not self-made representatives of a level professional playing field. Each of these tropes oversimplifies the reality, failing to draw attention to the fundamental *hybridity* of the professional workforce. This chapter explores how tech workers cultivate different roles, tasks, and barriers in their work and careers. We will examine how they align their critical worldviews with their embeddedness in capitalist organizations and juggle technical and communicative tasks in their jobs, and consider the barriers they have to overcome to "get in" and "get ahead" in their field. Finally, we will come to see how the institutional discourse codifies tech workers' professional roles and contributes to the dominant culture and power structures within the industry.

3.1 BROKERING NEEDS

Previously, we saw tech workers to be carriers of a renewed social critique of capitalism. They cultivate cultural and moral worldviews that are not hardwired to the post-Fordist ideals of a holistic market orientation as well as a naïve techno-solutionism. But how can we make sense of tech workers as social critics given their organizational embeddedness within the crown jewel of contemporary capitalism? How do such workers align their critical stance, as regards tech and society, with their labor market positions? And how did their path into tech corporations and start-ups enfold? To begin addressing this puzzle, let us turn to the case of Dalia, the UX designer from Berlin who had had previously worked as a financial consultant and whom me wet at the outset of this work. When I asked Dalia how she started working in tech, she told me:

> I graduated just after the global financial crisis in 2011. So, it was a little bit of a difficult time period for people in finance and accounting, but somehow I found a job. And then I started reading more about these emerging

> technologies, like blockchain, that emerged because of the global financial crisis. I was following it quite closely. [. . .] the reason why I wanted to learn more about technology and blockchain is because I saw them as avenues to improve people's lives and people's work experiences.

Like many of the tech workers I interviewed, Dalia mentions the 2008 financial crisis as an event that influenced her entry into the job market. She goes on to link her career in tech with her growing interest in technological innovations as well as her longing to improve society. Dalia narrates her entry into tech through the themes of reflexivity and social impact. For her, it is crucial to enfold social impact in her role, which she says she couldn't find in the two industries she had previously worked in: finance and consulting. A related field-entry narrative is provided by Sebastian, the Berlin tech worker who confidently self-classifies as a "tech worker." When I asked him how he got to working in tech, he told me:

> I actually wanted to go into banking [. . .] Because if you think about it, banking—like investment banking—is really one of the few areas where you apply, like, really high-end mathematics. And you get to build models that are, like, super interesting [. . .] So I did this. For six months, I did an internship, basically, in Paris, and I absolutely hated the industry. I hated pretty much everyone I worked with. They all seemed miserable to me. And I was like, I don't want to be this person in ten years. [. . .] And so that's why I think the fintech space was really good because, like, it's both challenging, but also, it's like the best of both worlds. So, like, you get the challenges of the finance industry, and the nice start-upy vibe that I like.

Similar to Dalia, Sebastian reflects on his venture into the world of finance as an experience that propelled him to reorient his career. Sebastian draws moral boundaries vis-à-vis the finance industry for its promotion of egoistic behavior, as he later specified. Thus, while tech workers can be quite critical of their own industry, it's important to note that they can be even more critical of adjacent industries. Dalia draws boundaries between tech and financial consulting, Sebastian between tech and investment banking. Finance and consultancy careers frequently serve as benchmarks in tech workers' professional evaluations. The two industries came up again and again as fields of comparison in terms of both career opportunities and moral ethics.

Sebastian reports feeling more comfortable in the fintech space he transitioned to: "I work in fincrime now. So, it's, like, financial crime prevention, and which is also a super interesting area to work in. So, I work in banking, but it's still trying to catch the bad guys. So, I guess for my conscience, it's good." Sebastian frames tech as a life world where he can pursue professional work that is both challenging and moral. The social code structuring this narrative is one of *balancing needs within oneself*. Importantly, this codex entails an economic mindset. Both Dalia and Sebastian do not consider a career beyond occupational fields with middle-class incomes. There are non-explicit limits to how far they are willing to go in making economic sacrifices to preserve their morals. It appears that tech workers are guided by the rationale of choosing the most moral (and challenging) industry that offers upper-middle-class jobs. Tech workers reach the conclusion that, for them, the world of tech provides the best compromise between economic and moral aspects. This consideration can be considered the most powerful *illusio* of the professional field of tech workers.

Drawing on Bourdieu, an *illusio*—understood as a particular belief that motivates participation in the actions and struggles of a field (Bourdieu 1990, 66)—is always also a *collusio* in the sense that it is a complicitly acknowledged implicit knowledge structure (Bourdieu and Wacquant 1992, 98).[1] The belief that tech work allows for economic and moral action binds distinct actors in the field of tech together; it must be displayed to access the microcosm. This is even true for tech workers that lean toward the affirmative ideal type, such as Ting, a UX designer who grew up in China. On the one hand, Ting seemed quite self-minded when she told me working in tech allows her to "benefit in my career path by learning and benefiting from a company." On the other hand, she also told me, "I think they [tech companies] are changing people's lives, making people's life easier. That is my goal, how to make people's life easier. [. . .] We're trying to slightly make people's lives better by building up a better product." Ting is drawn to the tech industry not just for its career opportunities but also because she considers it a habitat where she can pursue her goal of improving peoples' lives. Her moral-presentation does not strike me as purely strategical. Like her colleagues, Ting seeks to exact a combination of economic and moral capital through her work.

Interestingly, this hybrid longing for economic and moral capital does not translate into a strong loyalty toward employers. In examining their LinkedIn profiles, I noticed that most of my interviewees switch firms every few years—typically, they stay for around two to three years at a company. Only a handful of my fifty-two interviewees showed signs of a life-long marriage to their employers. In this sense, there does not appear to be a return to the modus operandi of the Fordist organizational self, who exchanged long-term loyalty for stable career prospects. Instead, tech workers pursue their career steps in ways similar to the entrepreneurial self, which stands for a high degree of flexibility and mobility. There is a hesitation among tech workers to commit. In a way, tech workers can be imagined as *moral surfers*: While they are not willing to take any wave that brings them to the top, they remain tied to the neoliberal schema of constantly evaluating professional opportunities and being ready to cut off their current occupational ties for the "right" wave.

Nevertheless, the subjectivation regime of the entrepreneurial self is transformed by tech workers' growing emphasis on integrating morals and ethics into their professional roles. A lack of consideration for others or for society more generally has become professionally devalorized. This was also the case for most of the freelancers I interviewed, though we might suspect freelance workers to be carriers of heightened forms of entrepreneurialism. The economically well-off freelancers, in particular, signaled clear detachment from holistic market orientations in their day-to-day professional practices. Levi, a UX designer from Berlin, even formulated a "personal manifesto." He told me he did not want to take on just any job anymore. Since he felt he had "the luxury to choose," he set up parameters for taking on jobs. The first parameter in Levi's manifesto is that his work should "create meaningful change," while the second parameter is "to work with talented people" and "deliver quality."

To understand the subjectivity of tech workers at a fundamental level, we must come to terms with their hybrid professional selfhood. They are neither opposed to the market nor do they cultivate an unconditional market orientation. Despite their critical stance toward the tech industry and capitalism in general, tech workers are not completely dissatisfied with the contemporary economic system. They are drawn in by some of its forces while nonetheless trying to keep their distance from it. Tech workers are on a quest to be *middle-class wealthy and morally worthy*. This is the deep implicit meaning system that I found to undergird the subjectivity of most of my interviewees.[2]

But let us dig deeper into how the quest to be middle-class wealthy and morally worthy relates to the day-to-day professional work of tech workers. Dalia, the UX designer who was unsatisfied with working as a financial consultant, was most explicit about this. In her account, we can make out a *homology* between worldviews and work practices. Just as she calibrates her life between economic and moral goals, so too does she attempt to balance business needs and user needs in her day-to-day work. Dalia describes this balancing of business and user needs as the most important skill to be a "good" UX designer:

> There are the stakeholders' needs, which are the business needs, there are the user needs, and then there are the engineering requirements or constraints. And you need to figure out how to balance all of these. And this is not something that you pick up from a bootcamp class at all, this is something that comes with experience. [. . .] And then you see the Venn diagram of stakeholder needs and user needs. And obviously, engineering constraints, that sweet spot, which is not very easy to get all the time, is where I try to design for.

Dalia does not just balance different needs within herself in terms of determining which industry and job to choose and stand by. Dalia is also oriented toward balancing needs in her day-to-day work. Her work practice follows the imperative to juggle different and even conflicting needs. She affirms that there are certain (capitalist) economic needs. The best work performance, for her, is one that finds "the sweet spot," the perfect balance between business and user needs. This hybrid subjectivity is not about code switching but about synthesizing distinct codes. Dalia takes the role of a professional mediator who must actively translate different needs. I repeatedly encountered this notion of *translating and balancing needs* within the workplace throughout my interviews. This ideal can also be clearly identified within Peter's sense of professionalism: "The role of the UX designer really is to understand the user, to bridge the gap between what the user needs to accomplish and what the company needs to accomplish." Such utterances allow us to interpret tech workers as almost diplomatic subjects. The ideal tech worker is imagined as one who creates the best possible compromise by understanding and then actively brokering different needs.

In analyzing their worldviews, it becomes evident that tech workers seldom conform to the image of soldier-like entrepreneurial selves who simply serve market forces. Rather, we can describe the professional backstage of digital capitalism as crowded by subjects who struggle with their inner self. With reference to science and technology studies (MacKenzie and Wajcman 1999), it can be assumed that this hybrid subjectivity influences what kind of innovations are pursued, and how, in a number of ways. In 2018, tech workers at Google even went so far as to protest their company's involvement in a Pentagon project that aimed to improve drone targeting through artificial intelligence. This example demonstrates the potential impact of tech workers' professional ethics. Under specific conditions, they can significantly disrupt the business operations of major tech companies. In general, though, the professional self of tech workers is oriented toward compromise within an economic order grounded in capitalist principles. These workers who discover patterns in data or design apps for a living usually don't want to risk their middle-class positions. They are willing to make moral sacrifices. Their professional role is undergirded by the aspiration to strike the best possible compromise between business and user needs. The social code of hybridity thus places limits on social critique. Tech workers secure their seats in the (upper) middle class at the price of having to continuously broker the distinct interests of others as well as the distinct interests that reside within themselves.

3.2 POST-NERDINESS

We have seen how depictions of tech workers as soldier-like, career-focused, and obsessed with disruption and riding tech unicorns falls short of grasping their genuine character. Let us turn now to the public trope that depicts tech workers as *nerds* (evident, for instance, in the HBO show *Silicon Valley*).

Through this cultural myth, tech workers are imagined as people who are good with computers but lacking in basic social capacities. They are framed as socially awkward subjects who are unskilled at interacting with people beyond their peer group of coders and hackers. This public image has a long tradition within accounts of computational knowledge workers (Kendall 1999). My interviews with tech workers, however, demonstrate that this image is, at the very least, outdated. While tech workers do indeed venerate technical expertise, they also value social skills highly and present themselves as communicative and empathic subjects. Based on my interviews, I argue that tech workers are carriers of a *post-nerdy subjectivity*. This self is fundamentally geared toward combining technical and social skills. The ideal tech worker subject is imagined as someone who synthesizes the expertise and curiosity of a nerd with the empathy and communicative capacities of a people person.[3]

The valorization of post-nerdiness among tech workers can be neatly illustrated through my interview with Rafael, a data scientist from Brazil working in the US. When I asked him what he considers to be the core qualities of a good data scientist, he identifies two skill sets, first elaborating on *technical skills*:

> There are sort of two ways of answering it. One of them is, like, you can like answer it by going down the technical skill route. Okay, if I'm going to be a data scientist, there's some sort of basic technological stack work you need to be able to do [. . .] you need some sort of data engineering skills to patch up the different data sources you're looking at. [. . .] And you need, you know, ways of analyzing it.

Sometimes data scientists also speak of these technical skills as the necessary "hard skills." Then, after elaborating on the necessary technical skills that a good data scientist must acquire, Rafael addresses the need for communicative capacities:

> I think another component, it has to do with communication really, like, how do you present the work you've come up with. [. . .] you have to present the work, even if it's a piece of software that nobody really sees directly, that piece of software must be presented and implemented.

Sociology has traditionally understood technical expertise and social competence as two different skill sets that are incorporated by distinct occupational actors (Riesman et al. 2001, 128). In my interviews, I find that tech workers construct an ideal of mastering both. While Rafael may be understood to attribute primary relevance to technical knowledge (which is rather typical for data scientists), he also argues that this skill set must be flanked with communicative skills. Being a technical genius is not enough: Tech work is considered not only a technical but also a social task. Tech workers want to be more than nerdy technology workers. Most of my interlocutors showed great concern for making their work understandable and being collaborative in order to fulfill the needs of different stakeholders. This can be interpreted as part of a strategic effort to define their professional identity against the potential future capabilities of machines. By emphasizing their communication and interpersonal skills, tech workers seem to be proactively preparing for a future where algorithms and AI could handle some of their technical tasks. Tech workers' codification as communicative subjects, however, is not only used to draw boundaries vis-à-vis nonhuman actors but also with regards to human subjects. Bernardo, another Brazilian data scientist (who migrated to Berlin, though), distinguishes data science from software engineering, which he considers to be a traditional discipline that lacks social skills:

> But I think this revolution of the data just started in the last ten or five years. And the software engineering discipline did not catch up with us with this [. . .] right now, we are developing algorithms that learn from the data, and that make decisions automatically based on the data. So, it's a completely different sport because you need to deal not only with the code, not only with customers, but with some aspects like social aspects, life care, fairness, life equity, like accountability, like bias.

Bernardo establishes boundaries vis-à-vis the more established profession of software engineering. The emphasis on social skills is widely articulated not only by data scientists but also by UX designers—indicating a cleavage between more novel and more established tech work professions (such as software engineering). While UX designers are more worried about being seen as purely creative or qualitative, rather than purely technical, they also frame social skills as key. Meiling, a UX designer from the US, for instance told me, "In general, I think good communication skills are very important. So that's, I think, the most important thing because most of your work is about communication with different people, different roles, different stakeholders." Another illustrative statement was provided by Tara, a Taiwanese UX designer working in the US. Tara explicitly connects the relevance of communication with the capacity to understand needs:

> I think most of the time, I talk to my business stakeholders and participants when possible. So, most of time actually [I spend] understanding these needs. I spend more time in the beginning of the whole project understanding the needs, and less time really coming up with the [design] variations—not because they're less important, just, I think, at this stage in my career, that part comes out quicker than it used to be. So kind of once you understand the need, then there is a picture in my head so I just need to put it down on paper.

Tara depicts the capability of understanding the needs of different stakeholders as a crucial element of good UX design. Interestingly, tech workers often codify empathy and the ability to function in a team as the keys to achieving this. Jonas, for instance, a UX designer from Germany, states, "I would say the best UX design moments are within a team. UX designers rarely exist as lonely people." It is important to note, though, that teams are not necessarily understood as level playing fields. As the UX designer David told me, "You are not working in an environment where the value of design is fully appreciated [. . .] So, design often gets kind of pushed into a corner towards the end of the creation of something to make it look good." A number of UX designers reported that their creative work is often less valued within project teams than the coding work of data scientists.

Nonetheless, what is clear is that both data scientists and UX designers valorize social habitats and place great emphasis on being communicative subjects. As Dalia states, tech workers are aware that they are "deemed to be nerds." They reflexively wish to avoid this role, cultivating a subjectivity that can be coined as *post-nerdy*. Whether you work back-end or front-end, whether you are a technical or a creative guru, getting recognition necessitates being able to demonstrate certain social skills. The days of simply producing good code without worrying about communication and networking are over, if they ever existed. The social code of post-nerdiness must be performed not just on social media platforms but also in the context of day-to-day professional interactions.

As stated, the categorization of tech workers as post-nerds should not lead us to diagnose the death of the social figure of the nerd. By conceptualizing tech workers as "post-nerds," I intend, instead, to point to a metamorphosis of this social figure of "the nerd" (Kendall 1999). Being obsessed with a new R coding package or a certain graphic design technique is still valued in the field. A good tech worker is supposed to be immersed in their work; introversion is not generally devalued. Yet, nerdiness is imagined as a character trait that must be flanked with communicative capacities. Being a hybrid post-nerd is constructed as the necessary social character for tech workers to be able to discover, understand, and communicate the needs of different stakeholders. Interestingly, this form of subjectivation also provides a new spin on forms of "techno-masculinity" (Poster 2013) among male tech workers. While masculine-coded technical competence remains an important element that continues to generate gender-based inequality, masculine behavior, in this sphere, can be seen to increasingly integrate forms of communicative and emotional labor that have traditionally been linked more with feminine roles (Hochschild 1979, 1983).

Apart from gender, the hybrid subjectivity of post-nerdiness also has implications for relations of race and class. Tech workers from the Global South will likely experience more difficulties with living up to the code of combining technical and social skills. As Amrute (2016, 67) has shown, tech workers from India and China are racialized in the German digital economy as subjects that excel at mathematical thinking but lack communicative capacities. Along these lines, Neely et al. (2023, 323) argue that white tech workers are 154 percent more likely to be promoted to leadership roles than Asian tech workers. Furthermore, studies from the Computer Supported Cooperative Work association also suggest that individuals from working-class backgrounds will have more difficulty living up to the code of post-nerdiness. Based on interviews with HR professionals at tech companies, Chua and Mazmanian (2020) found that a class bias existed in particular when applicants were evaluated with regard to nontechnical competences, such as "industrial fit," "organizational fit," and "individual fit."

By cultivating post-nerdiness, tech workers thus not only draw boundaries vis-à-vis other professions but also establish divisions within their own occupational segment and reinforce biased hiring processes. Any serious attempt to make tech organizations more inclusive must consider this insight. Successfully linking the character slates of the technician and the communicator is not an equally present characteristic across subjects in social space. In theoretical terms, my empirical research points to hybridity not as a general subjectivity (as depicted by some post-structuralists such as Deleuze 1994) but rather as an unequally distributed form of cultural capital. In the next section, this perspective will be further deployed when analyzing mechanisms of inclusive domination.

3.3 ENTRY BARRIERS AND RACIALIZATION

As mentioned, a major public trope about the tech industry is that it offers professional opportunities based purely on merit. All you need is a laptop and coding skills, the story goes, to break into tech and launch a successful career. I have already discussed how a number of social codes, including post-nerdiness, undermine this myth. In this section, I analyze how the professional field operates simultaneously through low and high entry barriers, both of which are tied to modes of domination. Firstly, I argue that the field of tech work is characterized by inclusive domination by means of low formalized educational boundaries that allow its professionals to mobilize expertise and social capital to dominate other professional groups within the competitive system of professions. This provides an important explanation for how the segment of tech work, though not protected by means of specialized degrees, still manages to professionalize. Secondly, I focus on tech workers from the Global South and show how they are subjectivated by means of a specific form of hybrid inclusion and exclusion that operates through racial regimes of inequality. Tech workers from the Global South face higher barriers when trying to get into the field as well as more obstacles when trying to get ahead. They find their expertise demanded while the rest of their selfhood is constrained by the notionally meritocratic structures of tech companies and start-ups. Both forms of inclusive domination, we will come to see, coexist and contribute, in different ways, to a socioeconomic system of differential inclusion.

Domination Through Low Barriers

To understand tech workers' inclusive domination with regard to low educational barriers, it's useful to return to basic insights within the sociology of professions. Parsons (1939) has already argued that established professions, such as doctors and lawyers, can be described as occupations with a high socioeconomic status that derives from their capacity to control privileged task jurisdictions (see also Muzio et al. 2011; Pfadenhauer 2003; Sheehan 2022). Securing one of the limited privileged task jurisdictions involves an array of exclusionary mechanisms that must be deployed within a competitive "system of professions" (Abbott 1988). Typically, these exclusionary mechanisms include the placement of high educational entry barriers (Freidson 1988). Requiring specific educational certificates allows professional fields to make sure that, for instance, not just any "charlatan" can claim to be a doctor—which could put the reputation of the profession at risk as well as create more competition among members of the guild. As Abbott argued, "[it] is common to professions to create rigid entry standards, coupling extensive education with several levels of examination prior to formal entry into the professions. This is part of a structure of control that seems utterly advantageous to professions" (Abbott 1988, 84).

In contrast to this traditional professionalization strategy, I found that no formalized educational entry barriers exist among the emergent professions of data science and UX design.[4] While it is generally necessary for tech workers to have some academic background, entry-level jobs are available to students from across basically all disciplines, with no standardized examinations in place.[5] This suggests that the credibility tactics tech workers use to establish occupational closure are increasingly drawn around cultural principles and belief systems rather than formal credentials controlled by professional associations. In this section, I want to draw attention to the fact that tech workers not only accept but cherish inclusive forms of gatekeeping. Tech workers valorize and propel a professional field formation that operates without formalized educational credentials. This is exemplified through a statement given by Peter, a senior UX designer from Germany, who problematizes specialized study programs for UX designers:

> I would say it [UX design] should be open to people from different backgrounds. When I'm in a position of hiring, I myself like to hire people that have studied something else or come from somewhere else because they bring a kind of value, whereas I feel like a lot of design programs these days are usually very centered around just learning skills. But they don't teach, you know, how to think, how to think critically for example [. . .] So my advice to anyone who wants to go into UX design or stuff like that is usually study something else and then do a half-year or one-year technical course to get the skills that you need to work.

Even though more and more academic programs are being established around UX design and human–computer interaction, Peter prefers ongoing UX designers to have backgrounds from outside of design. His hiring policy is at odds with how occupations traditionally professionalize. Importantly, not all of my interviewees were this skeptical of the academic institutionalization of UX design or data science. Younger tech workers, in particular, are more sympathetic toward, and more frequently trained within, specialized programs. However, almost all tech workers I met advocated for keeping the doors open to candidates without specialized training. Norris, for instance, a junior data scientist from the US, told me:

> So, I was taught at a boot camp, like a coding boot camp. It was a twelve-week, full-time, immersive program. It was not a formal, like, academic institution, which grants master's degrees or PhDs or anything like that. And I think that's unique. I think that there are a lot of industries where you do need that formal training ahead of time. But the bootcamp provided another way. So, I don't know if that is standardized. In fact, perhaps it's the opposite. I think that there are more diverse opportunities for people to move into the field of data analytics and data science.

Reflecting on his own path into the field, Norris links low formalized educational entry barriers with "opportunities." For him, the openness of data science to people from different academic backgrounds makes the field unique and distinguishes it from other fields. Norris values heterogeneity and inclusiveness when it comes to regulating access to his professional field. This valorization is frequent across the ideal types of the radical, reformist, and affirmative tech worker. Almost all my interviewees can be described as professionals who take pride in the fact that, as Cameron a UX designer from the US put it, "people don't have a typical path."

But how can we account for this untypical professional behavior? One interpretation lies in the "homology" (Bourdieu 1990, 248; Westheuser 2020) between views that value low formalized academic entry barriers and political worldviews that valorize diversity. There is a correspondence (if not to say an elective affinity) between tech workers' professionalization and their social critique. Given that formal entry barriers in the form of mandatory degrees and specialization constitute one of the most visible and explicit forms of exclusion, opposing these can be interpreted as a way for tech workers to establish authenticity and identity coherency across different issues and spheres (Lahire 2011).[6] Another factor that may account for the untypical professional valorization of unstandardized formal entry barriers can be located in the *emergent state* of many tech work professions. Both UX design and data science are relatively young professional fields within a rapidly changing industry.[7] Establishing fuzzy formal entry barriers may allow them to absorb expertise and social capital from different kinds of disciplines and networks. Their relatively high degree of openness likely outweighs the organizational and reputational risks for young professional fields. This interpretation connects to the research on "cultural omnivorousness" (Avnoon 2021; Childress et al. 2021; Friedman 2012; Peterson and Kern 1996), which has shown that inclusivity can generate advantages in struggles over recognition. By avoiding standardized academic entry barriers, tech workers are able not only to establish moral authenticity but also to build pipelines for dynamic transfers of expertise and social capital. This observation sheds a new light on the branding of tech as an inclusive habitat in the sense that genuine forms of educational inclusivity simultaneously constitute a strategy (without strategists) to exert domination within the competitive system of professions. Put simply, low formalized educational barriers prove to be a strength rather than a weakness in the emergent professional field of tech work.

Domination Through High Barriers

Let us now turn to debunking the myth that tech work is a meritocratic employment segment where talent and hard work prevail above all else. Despite the field's low formal academic barriers—a fact so often valorized by tech workers—there exist other kinds of tightly formalized entry barriers. This brings us to the second form of inclusive domination, which concerns *racialized relations of inequality* that manifest in subjugating visa regimes and experiences of precarious belongingness. In my interviews, I discovered that a number of international tech workers, especially those from the Global South, struggle with challenging and often less-than-transparent visa regimes in the US and Germany. While tech workers from the Global North find it relatively easily to enter the labor market in Germany or the US (least difficulties are experienced by tech workers from the countries of the European Union [EU]), others face much higher hurdles even though the demand for their labor power is high.

So even though tech work is indeed a highly international employment segment, which at first sight aligns with the myth about the tech industry as an inclusive habitat, its racialized entry barriers vary for different groups and its inner terrain is far from an equal level playing field. A paradoxical combination of exclusion and inclusion affects tech workers from the Global South, with manifold consequences even for those who manage to get in. I found that tech workers in the US and Germany who come from countries such as Brazil, India, or Peru often experience problems with getting promoted in their firms due to short-term contracts. In addition, they feel a higher degree of dependence on their companies, given that visas are generally tied to active job contracts. Arjun, an Indian data scientist working in Berlin, also reported negative consequences for his private life that derive from struggles with visa regimes. He told me about the great difficulties in getting his Indian spouse to join him in Berlin: "She was here during Corona time but only having a tourist visa, and then all flights got stopped. Then she had to go back in September now she's coming back on a Spouse Visa."

Thus, while tech professions appear more international than most others, such as law or medicine, and inclusive educational barriers are widely cherished and unite the professional segment of tech workers in struggles over status vis-à-vis other occupational fields, visa regimes generate a state of differential inclusion among them. Visa regimes are a central aspect of racialized relations of inequality in tech that are responsible for significant differences in professional opportunities. Tech workers from the Global South are simultaneously inside and outside of the field—they experience a subjugating form of inclusive domination that translates into experiences of *precarious belonginess*. As Pedro, a data scientist from Brazil working in Berlin, told me, "I feel the privilege I have in Brazil as well the prejudice [I face] in Europe." This statement is typical for individuals from the Global South who must juggle the benefits and costs generated through their migration. Furthermore, Pedro also reported different experiences of racialization across his new and old habitat: "For Europeans and Americans I am Latin; for Brazilians I am White." Pedro must not only balance the different needs and privileges of his transnational self; he must also deal with distinct racialized readings of his body. Moreover, being read as non-white comes with a number of disadvantages in both the US and Europe—while, of course, the ways these disadvantages play out differ across the US and Germany.

Thus, the tech industry is no more the meritocratic paradise many envision it to be than tech workers can be said to form a homogenous mass. The racialized subjectivation of tech workers from the Global South can be interpreted in the light of the theorizing of hybridity and ambivalence by Bhabha (1994), who addresses the split in the subjectivity of the colonized other. Bhabha highlights the mutuality of cultures in the colonial and postcolonial process, elaborating on how the colonized other can be understood as a hybrid, caught between their own cultural identity and the colonizer's cultural identity. Tech workers from the Global South who are laboring in the Global North can be understood as hybrids along these lines. Despite the many privileges they accrue from their so-called white-collar jobs in tech, they experience a duality of cultures and power relations that is absent for colleagues who do not have visa issues and do not constantly experience themselves read as not white. Tech workers from the Global South (as well as non-white tech workers) laboring in the Global North occupy an ambiguous position, neither fully integrated nor entirely excluded from the transcultural space of tech work, creating a paradoxical dynamic of inclusive domination.

Importantly, scholars of post- and de-colonial studies have underscored that processes of colonial racialization carry the potential for resistance (see Fanon 2008). As demonstrated by Amrute (2014, 2016), the specific hybridity of tech workers from the Global South can serve as a source for critical reflection and potential action. In her ethnographic study of Indian tech workers in Berlin, Amrute argues, "Indian programmers on short-term projects stressed the freedom of code to move across boundaries and do many things, a condition they contrasted with their own constrained mobility" (Amrute 2014, 107). This insight links to interests of science and technology studies scholars (e.g., Latour 2005; Wajcman 2018) who explore the nexus of technological affordances and social meaning systems. Scholars in this tradition highlight that material artefacts (including technology) and culture are no independent dimensions but intertwined dimensions. In a similar vein, Amrute demonstrates how the affordances of digital technologies, which tech workers take part in creating, spark reflections about their own positioning in global space. Furthermore, Amrute shows how this reflexivity connects to problematizations of power relations: "I believe that often, by using code to think about the conditions of their own lives, programmers expose a neoliberal model of achievement that simultaneously rewards expertise wherever it is found and sequesters subjects within the concrete realities of embodiment and place in the name of rationalizing expertise" (Amrute 2014, 110).

Drawing on this perspective, I hold that it is crucial to highlight the subjugation as well as the potential for emancipation entailed in the hybridization of tech workers. On the one hand, tech workers from the Global South must indeed grapple with an economic system that aims to integrate their labor power but does so in the context of a system of differential access that is interwoven with racialized relations of domination. Yet, on the other hand, tech workers from the Global South are no passive agents but rather embedded actors with critical and reflexive cultures related to their specific subjectivation. It is crucial to envision their hybrid experience not only as one of subjugation but also as one of potential empowerment. Processes of domination and emancipation are summoned simultaneously through the multiple forms of hybrid subjectivation of tech workers. By rejecting simplistic narratives to acknowledge tech workers' fundamental hybridity, we can better understand their social embeddedness and the interconnected complex mix of challenges as well as opportunities for emancipation.

3.4 PROFESSIONAL ROLES AND INSTITUTIONAL DISCOURSE

From the interviews, I have reconstructed three central ways in which the professional roles of tech workers are structured by hybridity: (1) they flank their social critique with the professional code of brokering different needs, (2) they combine technical and social capacities in a post-nerdy fashion, and (3) they are characterized by two different forms of inclusionary domination that manifest in a paradoxical copresence of high and low entry barriers. Now I turn to the question of the extent to which these forms of subjectivation correspond with the institutional discursive constructions of tech workers. Is there a correspondence between tech workers' professional subjectivation and their discursive interpellation in study programs and job ads?

Regarding the code of balancing needs at work in the pursuit of being middle-class wealthy and morally worthy, I find a modified correspondence in the discourse. Both job ads and study programs frame the tech field as a habitat of opportunities for simultaneously achieving career success and improving the world. For instance, the German internet company HelloFresh, which delivers prepared food boxes, promotes itself as a firm where "we will encourage you to make an immediate impact in your area of work as well as empower you to grow your career with us" (HelloFresh, GER, DS). Another German tech company, Delivery Hero, demands, and offers the opportunity to, "ensure high business and user impact" (Delivery Hero, GER, UX). The same mantra is deployed by US tech companies: "We follow a simple but vital premise in the UX group: 'Focus on the user and all else will follow.' We're interested in our users and strive to learn everything we can about their behaviors, attitudes and emotions to help define the products and experiences we create" (Google, US, UX). Notably, though, the economic field does not classify business and user needs as being in conflict. The two need structures are generally imagined as *harmonious rather than conflicting*. Hence, while tech workers spoke at length about the necessity to balance distinct and conflictive needs, this professional skill is not called upon in the discourse. By sidestepping the trade-offs between ethical concerns and business interests, tech firms only partially respond to tech workers' professional codex.

Another difference can be located in the heightened *revolutionary rhetoric* of the discursivations by the economic fields. This can be illustrated through the following job ad by Amazon: "We're working hard, having fun, and making history. Come join our team! You will have an enormous opportunity to impact the customer experience, design, architecture, and implementation of a cutting-edge product used every day by people you know" (Amazon, US, DS). Across national habitats, across professions, and across different types of companies, the discourse of the economic field articulates the promise of economic opportunity and world improvement not through a rhetoric of compromise but rather within a hyper-rhetoric of revolution. At start-ups, in particular, this rhetoric is interwoven with the figure of the entrepreneurial self: "Our design 'team' is one person. You can move as fast as you want. Plus, given our tremendous traction, your very first strokes will be seen by millions. This combination is rare and quite satisfying" (Padlet, US, UX).[8] This contrasts sharply with tech workers' self-understandings in the sense that it frames them as bold innovators who move fast and break things rather than as mediators balancing diverse needs.

Somewhat surprisingly, this revolutionary framing is echoed in the academic field. The language used by universities in the context of their study programs veers quite close to the economic repertoires. The University of Virginia, for instance, interpellates tech workers as "leaders" and "transformers":

> We are creating leaders in the cutting-edge field of data science and equipping our graduates with the technical and practical skills to transform the modern world. Graduates from our Master of Science in Data Science (MSDS) program are changing business, medicine, engineering, technology, and public policy, just to name a few fields. You will leave us as a data scientist poised to tackle the world's biggest problems. (Virginia, US, DS)

German universities tend to be somewhat more reserved in their cultural repertoires. It's possible that this is the case because their programs are generally tuition-free, so they are under less pressure to brand their offers. However, even some German universities join the chorus of promoting tech work as a revolutionary and entrepreneurial task area. Consequently, the academic fields cannot be understood as a space that echoes the self-codification of tech workers as employees who are oriented toward balancing needs. While it's important to note that the academic fields are not free of moral codifications (in Germany, tech workers are constructed as protectors of data privacy and, in the US, rather as challengers of data bias), the figure of the entrepreneurial self remains much more prevalent in their discourse than within tech workers' self-understandings.

But what about post-nerdiness? At first sight, we find a striking correspondence with regard to this code. Across the economic and academic fields, tech workers are also codified as both technicians and communicators. At times, this sentiment even manifests within one single sentence: "We're looking for someone who has the technical skills to surface insights quickly, and the interpersonal skills to communicate those insights in a way that persuades action" (DoorDash 2019a, US, DS). But what specific dispositions and schemas constitute this discursive construction of tech workers as technicians, on the one hand, and communicators, on the other? We can make out a *generalist* set of methodologies and tools that are discursively assembled in relation to the figure of the technician. Within study programs, we find curricula that ceaselessly emphasize the interdisciplinary technical training of ongoing tech workers: "The BS in User Experience Design (UxD) degree is an interdisciplinary, hands-on program that integrates creative and technical disciplines such as graphic design, human-computer interaction, and information technology" (DePaul University 2020, US, UX). Study programs promise an education that prepares students for professional data and design work through a pluralistic technical skill set. This generalist technical imagination is also present within the economic fields. It manifests, for instance, in demands for tech workers to work end to end: "Own end-to-end user journey and all aspects of design execution from vision, ideation, to prototyping, to user testing, to final production" (Coinbase, US, UX).

Notably, though, the discursive construction of tech workers as technicians also frequently entails the theme of *specialism*. In both the academic and economic fields, tech workers are interpellated as subjects who should acquire expert and detailed knowledge about one particular methodology or technology. Duke University, for instance, imagines its graduates to acquire broad experience while also specializing: "Experience the full range of the data science ecosystem and graduate as an expert in at least one analytical approach or branch of technology" (Duke University, US, DS). In the economic field, specialization appears as a theme, in particular, by way of demanding that applicants acquire domain knowledge in order to "create a highly differentiated and personalized experience for consumers" (The Yes Product Designer, US, UX). The combination of generalism and specialism is considered necessary as a means of discovering and commodifying the singular needs of customers.[9]

Let us now explore in more detail how the discourse constructs the hybrid counterpart of these technical capacities: namely *social* capacities, linked to the figure of the communicator. Similar to the self-understandings of tech workers, the discourse is concerned with negating associations of tech workers with pure technical "nerdiness." Communication is presented as an integral part of curricula, especially at elite universities, and justified via the perception that "clearly communicating problems, ideas, data, analysis approaches, results, and recommendations for action are vital for career success in technology and science" (Georgetown 2019).[10] Tech workers are explicitly curated as work subjects with social and empathic skills—capabilities that are the object of training through various measures including special seminars, group work sessions, and team project competitions. Actors with technical expertise are no longer considered mutually exclusive from those with communicative abilities. Tech workers are interpellated as the best of both worlds. Put bluntly, the ideal tech worker is expected to combine the technical virtues of a Steve Wozniak with the communicative virtues of a Steve Jobs.

Within the economic field, the interpellation of tech workers as post-nerds is even taken a step further. This is exemplified in a job ad by the tech company Microsoft, which describes their ideal candidate as being a "leader" and a "no-ego" character:

> The ideal candidate is a no-ego, inclusive design leader who is passionate about working across a complex design landscape, is human-centered, tenacious, and able to bring clarity through design that moves people and technology. In this role, you will drive cross-discipline collaboration with Program Managers, Engineers, Data Scientists, Design, and Research to create world-class experiences that inspire. You are outcome-focused, utilizing storytelling, craftsmanship, and partnership along with your insight to develop creative and pragmatic solutions. Candidates must demonstrate the ability to execute high-quality design work within time and technical constraints, amid competing priorities and multiple stakeholders. As a Senior Designer, you advocate for the value of design excellence across all disciplines at Microsoft and are committed to building a vibrant and diverse design community. (Microsoft, US, UX)

This job extract places an additional layer of hybridity on top of the figure of the post-nerd. It imagines the ideal tech worker as someone who, based on their technical and social capacities, generates output and performs leadership in an inclusive fashion. It suggests that leadership ought to be cultivated in a less directly hierarchical fashion (which is in contrast with Steve Jobs's more autocratic leadership style back in the day). In a sense, this connects to Foucault's concept of the shepherd (Foucault 1982; Waring and Latif 2018). Marking a manifestation of pastoral power, tech workers are interpellated as shepherding post-nerds who can subtly guide various self-governed subjects within the project teams of the digital economy. Within the discourse, tech workers' ideal of balancing different interests is thus channeled into a corporate culture where power and authority operate in more latent forms.

Finally, let us turn to the multifaceted phenomenon of inclusive domination and its echoes in the discourse of the academic and economic fields. With regard to racialized relations of inequality, unsurprisingly, there is not much of an echo. It is only through academic courses on "data ethics" as well as job ads' rather standardized articulations of openness to applications from individuals from marginalized social backgrounds that race is thematized at all. The existence of subjugating visa regimes and the particular challenges of belongingness for tech workers from the Global South are not addressed. With regard to the manifestation of inclusive domination through low educational barriers to entry, though, we can find a more pronounced (if divided) echo. The job ads show a strong correspondence with the appreciation of diverse academic backgrounds. No job ad demands that tech workers come from a specific academic field let alone hold a specific UX design or data science degree. Job ads for UX designers are particularly unspecific here. Regarding data science job ads, most companies simply state that they are looking for applicants with a quantitative background: "The ideal candidate will have a background in a quantitative or technical field, will have experience working with large datasets, and will have some experience in data-driven decision making" (DoorDash, US, DS). In the academic field, on the other hand, there is less appreciation of fuzzy entry barriers, especially in data science. While universities are open to accepting students from a variety of undergraduate programs for their master's degree programs in data science, they often require a "foundational knowledge" (Virginia, DS, US) in specific subjects, generally including statistics, mathematics, and computer science. Unsurprisingly, universities also tie specialized study programs to better preparation for the respective job as a tech worker: "While the curriculum and work for this project are demanding, it provides our students with valuable user-centered research, iterative designs and product development experience before they step outside of our halls and back into the business world" (Carnegie Mellon, UX, US).

The analysis of the discursive interpellation of tech workers against the background of their self-understandings once again reveals a relationship of partial correspondence. While the discourse utilizes the theme of balancing needs, the discourse frames this hybridity as free of tension. Furthermore, the discourse constructs tech workers as post-nerds in a more heightened and differentiated way than do the workers themselves. At the same time, the discourse analysis revealed that racialized relations are barely thematized, while the question of academic entry barriers shows a divided echo in terms of the discourses of the economic and academic fields. A qualitative gap exists between how tech workers codify themselves and how the institutional discourse interpellates their subjectivity on these issues. This supports the view that we mischaracterize tech workers when we see them as mere soldiers of the entrepreneurial class whose hearts and minds straightforwardly align with the corporate cultures of their firms. At the same time, we have seen key ways in which the discourse strategically echoes tech workers' social codes in an attempt to co-opt and appropriate these.

4 AMBIVALENT LIFESTYLES

Tech's urban takeover extends far beyond San Francisco. All over the world, we are witnessing a transformation of city space. In Berlin, for instance, the central district of Mitte is being reshaped by a myriad of start-ups. Nearly every street seems to be home to an aspiring "unicorn." Crossing the historic site of the Hackescher Markt, we find that SAP, the largest German tech corporation, has taken over the building that was once the headquarters of the German Communist Party's central committee. Inside these walls, people still discuss "transformation" and "disruption," but now of a very different sort. Economic dominance has supplanted political dominance. Digital capitalism now reigns, and it is reshaping the social fabric of the city. Tech companies' takeover of urban and social space goes hand in hand with the influx of new residents. Digital professionals are reshaping both Berlin and San Francisco, cities that were once imprinted by creatives, academics, and political professionals. The contemporary technological transformation of urban space manifests not only in omnipresent electric scooters or overworked food delivery couriers but also in a new crowd of middle-class residents. Tech workers are the labor aristocracy of the contemporary city, equipped to absorb, though with increasing difficulty, the skyrocketing rents and real estate prices. But what characterizes the lifestyles of these newcomers? What kinds of cultural tastes and leisure activities do they bring to the city?

Building on the previous examination of tech workers' worldviews and professional roles, this chapter turns to lifestyle as an important component in analyzing their subjectivities.[1] To gain a thick account of what makes them tick, we also need to understand their everyday cultural preferences. Drawing on my interviews, I identify three typical codes that characterize the cultural tastes and leisure activities of tech workers. First, a code of *comfortable exploring* shapes their lifestyles. While tech workers remain attached to post-Fordist romantic values of self-discovery and self-actualization, they simultaneously seek economic and moral comfort. They strive to explore new sites, situations, and subjectivities without experiencing economic precarity or moral destabilization. Tech workers stay near the edge without stepping over it. Striking a new Faustian pact, they accept the high demands of a hyper-globalizing social order in exchange for personal stability. The second lifestyle code that emerges is an emphasis on *ordinariness*. Though tech workers are guided by a romantic urge to explore new habitats, they accompany this modus operandi with a valorization of ordinary, everyday tastes and leisure activities. If they engage in exclusive lifestyle activities at times, most tech workers actively present themselves as down-to-earth. This cultural positioning is highly ambivalent as it not only holds the potential to level or blur cultural boundaries but also allows tech workers to sidestep appearances of elitism even as they ascend beyond a declining middle class. A third pattern characterizing the leisure activities of tech workers centers on a code of *mindfulness*. My interviews reveal how tech workers cultivate a range of technologies of the self in order to counter accelerationist tendencies and feeling of disconnection. This social code holds promise for a redifferentiation of work and life and a move toward new forms of care. However, mindfulness is also double-edged: Its current enactment does not break with the epistemic doxa of self-responsibility. A close analysis of tech workers' lifestyles exposes their everyday frustrations with neoliberalism but also the ways in which their evolving social codes are still rooted in many of its fundamental principles.

The fourth part of the chapter will again contrast the self-understandings of tech workers reconstructed from my interviews with my findings from the discourse analysis of relevant job ads and study programs. My overarching argument here is that tech workers' self-presentations as comfortable explorers, ordinary folks, and mindful selves find only a partial echo in the economic and academic discourse. There is no general correspondence across the two domains of knowledge. Instead, only select codes are echoed, and often in appropriative ways. This further supports the observation that economic firms and universities respond to the social codes of tech workers only insofar as these can be aligned with their organizational interests and facilitate processes of commodification.

4.1 COMFORTABLE EXPLORING

The notion of *exploring* implies a social figure who sacrifices a comfortable existence for the chance to discover something new. This figure, which constitutes a central component of Western myths (Yusoff 2012), played a key role in the post-Fordist world of work. The allure of self-exploration led many middle-class professionals to willingly trade diminished economic and institutional security for opportunities to explore their individuality. Boltanski and Chiapello (2018) theorized this process as the capitalist appropriation of the "artistic critique."[2] In the lifestyles of contemporary tech workers, the figure of the explorer surfaces again, but with a new social character. This figure adheres to the code of *comfortable exploring*. Bluntly put, tech workers' lifestyles are calibrated toward sheltered living at the edge of new developments. Their private life oscillates around fusing exploration with economic and moral comfort. This new spin on the post-Fordist lifestyle is illustrated in a statement by Jennifer, a data scientist working at a big tech company in the Bay Area. She reflects on her lifestyles in a two-fold way:

> I'm very lucky that I'm in a field that pays well, so I'm very comfortable and can live the life I want. [. . .] I don't have to worry about finances. I mean, I'm pretty set up as far as, you know, retirement goes. I got a good retirement plan. I have plenty of money. I mean, I can live by myself which I've always wanted to do. Not a lot of people my age in this area can live by themselves. [. . .] And then as far as traveling goes, you know, in three months, I'll be out of my lease. So, I might, you know, just travel for three months and work remotely. So, I just have that flexibility of being able to do whatever I want to do with my life outside of work.

Jennifer longs for autonomy outside of work. Literally, she desires to determine the principles (nomos) of her private life on her own (auto). It is also clear that Jennifer links autonomy with economic comfort. For Jennifer, being independent requires a certain income and a retirement plan. Retirement plans may have been difficult to imagine for explorative white-collar workers in their late twenties under post-Fordism, and they were surely unsexy for the counterculturalists who crowded the Bay Area a few decades ago (Turner 2008). But it is not so for contemporary tech workers. Material security now and in the future matters. The post-Fordist pact of trading economic security in return for creative missions, self-discovery, or other acts of exploration is being questioned by many of my interviewees. They still want to explore the world as well as themselves, but not at the cost of economic comfort.

Jennifer also shows the high value she places on traveling. Jennifer longs to live in new places and discover new people. Already moving away from Ohio to go to college in Virginia marked a positive experience for her: "So college was a great experience. I met a lot of people from different backgrounds, which was cool because Ohio was not so diverse." Such a valorization of traveling and meeting new people is shared by almost all the tech workers I interviewed. Also, tech workers from the Global South with temporary visas told me they do it as far it is possible. In general, tech workers are rather positive about globalization. In contrast to current macroeconomic developments where protectionism or "reshoring" are gaining hold, tech workers are in favor of more globalization rather than less. They typically consider the globalization of the economy not as a threat but as an opportunity. However, they are also keen to point out the importance of living a cosmopolitan lifestyle with due regard to economic security. Roger, for instance, an Australian UX designer working in Berlin, told me that working in tech allows him to "have enough savings that if anything was to go wrong, and I needed to get back to Australia, I could do that."

Tech workers want to explore the world without risk. They combine romantic and cosmopolitan ideals with a valorization of comfort. Living spontaneously and without a safety net is not for them. Furthermore, tech workers seek to bring their lifestyles into harmony with some of their moral principles. Regarding the social code of comfortable exploring, tech workers are most keen about aligning this lifestyle with principles of *the right ecological way of life*. A lot of my interviewees are aware that their intensive traveling conflicts with their displayed environmental consciousness. Moritz, for instance, a data scientist from Berlin, demonstrates this when reflecting on his vacations:

> So, when I see flights, I see hey, you can balance your CO_2 emissions now if you donate two euros. I'm not fully sure how this works but I believe it, maybe because I want to believe it. So, I do this basically every time. [. . .] And then, smaller restrictions, like within Germany, I don't fly anymore, these kinds of things. If I go to Poland or Amsterdam, for example, I take the train even though it's a little uncomfortable. So, I'm trying to find a balance of not doing things if it's still feasible.

Like many of my interviewees, Moritz is concerned about aligning exploration with *moral comfort*. He seeks to balance or reduce the CO_2 emissions he incurs by traveling. He is willing to accept some extra costs (for compensation products whose effect he is not sure about) or levels of discomfort in return for eco-moral comfort about his cosmopolitan lifestyle. This reveals that the modus operandi of *balancing different needs* shapes not only the professional but also the private lives of tech workers. At the same time, the statement demonstrates that an ecological cosmopolitanism requires a certain amount of economic capital as well as the ability to craft a moral narrative around one's lifestyle choices.

Notably, the exploratory nature of tech workers' lifestyles extends beyond travel. Moritz shows characteristics of the comfortable explorer with regard even to daily activities: "Checking out new restaurants is probably one of my biggest hobbies [. . .] everything food related and sports related." Indeed, the code of comfortable exploring surfaces very explicitly in tech workers' accounts of their sports activities, which are structured by an ethic of *exploring one's body through controlled quantification*. One stereotype that appears to hold is the popularity of self-tracking among tech workers. A substantial amount of my interviewees wore smart watches and FitBits (an electronic bracelet that allows you to track bodily activities). These devices are particularly well suited to quantify the rather individualistic sports activities most of my interviewees preferred, such as running, climbing, undertaking hikes, or going to the gym. Ruth, a data scientist working in Berlin, told me:

> Yeah, I just like exercising. Running is like the thing for me that makes me the most tired [. . .] And it's just fun to track [. . .] [I]t's nice because you can again, like set yourself goals and targets and you can measure everything. And that's nice. Like you can say, I want to hit this time for my marathon, or I want to run this amount of kilometers a week.

Ruth likes running not only because it provides a comforting tiredness but also because it allows her to measure herself; she wants to explore her bodily performances. There seems to be a will for knowledge about her own body. Interestingly, Ruth herself draws a connection between this mode of conducting sport and the way she likes to work as a data scientist:

> I like to do lists, and I like checking things off. And I like merging pull requests. And these things that are, I would say, like, define goals that you work towards, and that you can meet and achieve. And then like, it's like running, right? If you run ten kilometers, like [you get one] kilometer after [the other] kilometer done. And of course, you can build those things in your academic research, but it's less definable.

Ruth, who had once considered an academic career, seeks clearly definable achievements in work and sport alike. She reflexively identifies a correspondence between how she tracks her running and how she works. In a way, this homologous thinking can be considered as quasi-sociological in its form (Boltanski and Chiapello 2018; Giddens 1991). In general, I found the degree of self-reflection among tech workers striking. Many of my interlocutors not only observe their behavior closely for patterns but also interpret these patterns in elaborate ways.[3]

From another perspective, I interpret Ruth's statement as indicating a longing for certainty about what is morally right. By meticulously tracking her activities, she seeks to impose a sense of order and control over her life, exploring her physical experiences in ways that feel concrete and comprehensible. The phenomenon of self-quantification has been widely examined, but here, it is crucial to consider its dual aspects of self-optimization and normative control (see also Neff and Nafus, 2016, 24). Among tech workers, there exists a deep longing to establish some certainty within a world that is perceived as uncertain. It appears that tech workers are deploying self-control techniques with the dual goal of improving their bodily capital and of attaining higher levels of stability in day-to-day life. Thus, while the neoliberal principles of self-optimization and self-control remain an important culture of subjectivation among tech workers, self-quantification as technology of the self (understood here in a very literal sense) is also deployed to establish some certainty over a world experienced as increasingly in flux.

Last but not least, comfortable exploring can be seen in tech workers' willingness to accept my interview requests. Recruiting participants was easier than I had expected, a fact I interpret as the result of two key factors. Firstly, inviting individuals for interviews via Zoom lowered the threshold. All my interlocutors were well acquainted with the software and likely preferred the reduction in logistical efforts compared to in-person interviews. Secondly and perhaps even more importantly, I sensed that many of my participants agreed to be interviewed out of a dual sense of curiosity and reciprocity. After the interview finished and I had turned off the recording, a lot of them asked me questions about my preliminary findings. Notwithstanding mere politeness, it appeared to me that they genuinely wanted to explore new perspectives about their own occupational segment—an economic grouping that many of them felt was not being noticed by people outside of tech (especially in Germany). Furthermore, some tech workers I talked with seemed to have agreed to be interviewed not just out of curiosity but also out of a sense of moral duty, or as one put it, "to give back." These tech workers (mostly UX designers) told me they themselves heavily rely on interviews in their work and therefore felt it was right to take the time to participate. Thus, it appears that interview participation was often perceived as a manageable act allowing for exploration and/or a sense of reciprocity.

4.2 ORDINARINESS

In his seminal work on "Distinction" (1984), Pierre Bourdieu shows how lifestyle preferences not only represent class positions but actively (re)produce these. In Bourdieusian terms, the presentation of the self as a comfortable explorer generates class reproduction insofar as its enactment requires relatively high volumes of economic and cultural capital. However, it is important to note that the lifestyle preferences of tech workers are not always tied to substantial economic or cultural resources. Through conversations about their leisure activities and tastes, I discovered that alongside their interest in comfortable exploring, they are invested in maintaining an image of ordinariness. Across the radical, reformist, and affirmative types of tech workers, there exists a codex of cultivating exploration without appearing elitist. Despite average salaries of $85,000 (73,000 euros) in Germany and $162,000 in the US, most tech workers I interviewed were keen to emphasize their appreciation of unpretentious leisure activities and tastes. This is illustrated in a statement by Josh, a UX designer from the US, who told me about his ideal night on a weekend:

> I like to play board games and I play a little bit of video games. Also, I watch TV. I like to have parties, you know, make a meal, host people, play board games, play drinking games, go out. I like to eat out, you know, go to bars. The ideal night for me, it's like, make a big meal, have some friends over, eat, play some board games, have a bunch more people come over, get really drunk, go to the bars, see some great music. That's a great night for me.

Josh, who recently moved back into his hometown area around Chicago, does not strive to cultivate an extraordinary lifestyle. There is no sense of singularity in the presentation of his leisure activities. What makes a perfect night for Josh is a mixture of community, entertainment, and socially accepted drugs—elements that do not mount a challenge to the dominant culture and do not require high levels of cultural or economic capital. Of course, some of the listed activities are not entirely inclusive. They require a certain amount of discretionary income, which only increases when they are cultivated in the setting of pricy urban habitats.

We witness a similar signaling of ordinariness in the case of Rachel, the data scientist from Austin we met at the outset of this study. When I spoke with Rachel about her leisure activities, she told me, "I like to run, you know—hang out with friends and family and go to concerts. You know, I have a little dog. And I like to kind of take my dog on walks and go to, like, coffee shops and things like that." As was the case with Josh, Rachel foregrounds leisure activities that require some economic expense, but they are not calibrated toward cultural exclusivism. Of course, I did encounter tech workers who leaned more toward exclusive cultural pursuits, such as attending art exhibitions. However, the majority of those I interviewed appeared keen to emphasize their ordinariness.

But what makes this interesting? Is it really so surprising to find middle-class individuals reporting ordinary lifestyles? At first glance, perhaps not. Sociologically speaking, ordinariness has long been observed among the middle class (Savage et al. 2001). However, this lifestyle becomes striking when it is adopted by upper-middle-class individuals with high cultural capital, a segment cultural sociologists typically associate with rather exclusive, singular tastes and lifestyles (Bourdieu 1984; Reckwitz 2020). The majority of the tech workers I met told a different story (what their everyday lifestyles look like in practice is another question). Even though my interviewees were economically well-off and all but one held university degrees, often from prestigious universities, they deliberately *foreground* a lifestyle of ordinariness. It is this foregrounding that sets them apart from so-called cultural omnivores who combine low-brow and high-brow cultural activities in their lifestyles (Peterson and Kern 1996). The emphasis tech workers place on ordinariness, in combination with their high resources, becomes a signature move within the upper middle class; it translates into a relatively distinctive conduct of life. Contrary to theories that paint the well-educated middle class as indulging in refined and exclusive lifestyles, I find that the digital middle class is more concerned with cultivating relatability and projecting a down-to-earth image.

At times, my interviewees even signaled awareness that singularistic or unusual leisure activities are expected of them on account of their class position. Indeed, tech workers are quite reflective when it comes to the connections between lifestyles and social status. They recognize the trend of singularization within the upper-middle-class while positioning themselves to the side of this development. This came to light when my interlocutors described their lifestyles as "nothing crazy" (Roger, UX, GER) or "not too unusual" (Lisa, UX, GER). Importantly, these comments were made without a hint of low self-worth. Tech workers generally stood by their ordinary leisure activities with confidence. Notably, the perceived need to somehow justify or explain one's ordinariness was more prevalent in Germany, which is consistent with Lamont's finding that ordinariness is more appreciated in the US than in Europe (Lamont 1992). I nevertheless hold that the code of ordinariness among tech workers in the US is significant: The degree to which tech workers foreground low-brow lifestyle activities is very high and arguably distinguishes them from other upper-middle-class fractions, such as consultants or bankers.

The valorization of ordinariness can also be witnessed in tech workers' conduct of romantic relationships. Most of my interviewees were either embedded in long-term relationships or they told me they would like to be in such relationships. Tech workers generally do not appear to engage on the partner market as actors who are constantly trying to improve their match. While they do not want to commit to one firm for long-term employment (many feel that the "learning curve" drastically drops after three years), they typically do want to commit to one person for long-term romantic relationships. When I asked my interlocutors what values were important for them in a relationship, they often referred to principles of "loyalty" or "trust" rather than "creativity" or "inspiration." These orientations can be interpreted as limiting the intrusion of market logics into the sphere of romantic relationships as described by Eva Illouz (2007). When it comes to intimacy, devotion is valued over discovery. Tech workers' orientation toward loyalty and trust in their romantic lives may be a response to market pressures in the sense that ordinary relationships function as a private safety net in an increasingly deregulated society.

Turning from personal relationships to tastes, the codex of ordinariness can be located in tech workers' preferred choice of a "no-collar dress code" (Ross 2003). In the interviews, tech workers often wore rather inconspicuous tops, such as plain t-shirts or hoodies (which is not to say that they looked cheap). My own style, which usually followed the academic convention of pairing a shirt with a wool pullover, made me appear rather overdressed by comparison. Of course, for all their simplicity, tech workers' inconspicuous clothes still function as symbolic goods in their respective field. The dress code of tech workers latently signals that they do not grant much relevance to appearance but use their time to think about other issues. This also seems to be the case with the choice of virtual backgrounds on the zoom platform. When at home, most of my interviewees were sitting in front of a seemingly neutral background (very often a white wall). Unlike academics, not a few of whom favor massive bookshelves as backgrounds, tech workers rarely display objectified cultural capital from their field when appearing on video calls.

The widespread display of ordinariness across leisure activities and tastes makes it a crucial factor in the nexus of lifestyles and distinction. Despite exclusive tastes among tech workers and their preference for pricy metropolitan living places, my interviewees were keen to cultivate and convey down-to-earthness. This can be considered a moral orientation in the sense that it builds on a cultural valorization of inclusivity. At first sight, ordinariness therefore apparently contributes to leveling cultural boundaries vis-à-vis lower status groups. Drawing on Friedman and Reeves (2020), however, the logic of ordinariness that permeates the tech industry must be considered not only a moral phenomenon but also as a strategy (without strategists) for establishing cultural legitimacy. Friedman and Reeves have demonstrated how the British elite increasingly relies on ordinary modes of distinction to secure their position as well as the growth of their economic capital. My study indicates that this mode of distinction is also present in the upper-middle-class. Without considering every element of the subjectivity of tech workers as strategic, I hold that the presentation of an ordinary self allows them to avoid becoming a target of critique. Cultivating ordinariness contributes to the fact that unlike financial professionals (Neckel et al. 2018), tech workers are not so much the target of public critiques even though their industry is under similar public scrutiny. Ordinariness places an inconspicuous cloak around tech workers, allowing them to pull away economically from a generally declining middle class without making a scene.

All this said, processes of subjectivation are never black or white. There is potential for both domination and emancipation through the cultivation of ordinariness. There is a space of possibility that may be navigated in pursuit of emancipation through a reduction of lifestyle boundaries as well as a reproduction of relations of domination through even more subtle and inconspicuous cultural mechanisms that allow for intensified economic inequalities. As Hall and Du Gay (1996, 4) have highlighted, the self always emerges out a play of multiple and contradictory modalities of power.

4.3 MINDFULNESS

Another code through which tech workers transform, but do not completely break with, neoliberal forms of subjectivity is the lifestyle of *mindfulness*. Importantly, tech workers' adoption of mindfulness goes beyond common understandings of the phenomenon as centering on yoga and meditation. It must be understood as a larger way of life with multiple components, centered around *care for mental and ecological capacities* (Leggett 2021; Purser 2019). The care for mental capacities manifests prominently in tech workers' insistence on a differentiation between work and life. As mentioned in the section on the return of social critique, most tech workers are skeptical of an entrepreneurial way of life that dissolves the boundary between work and life. They seek to keep the two spheres separate. Many of my interviewees chose to work in the tech industry because of its seemingly better work–life balance compared to finance, consulting, or academia. Almost all tech workers I interviewed were able to tell me how many hours they worked per week (on average, fifty in the US and forty in Germany). In opposition to a subject that is always available and never really logs off, tech workers revalorize aspects of the work regime of the organizational self. Consequently, the introduction of widespread remote work in the context of the Covid-19 pandemic was carefully adapted by most of my interviewees. For instance, Sophie, a data scientist from the US, told me:

> So, one thing that I personally did was, I moved from a one bedroom to a two bedroom. Because I had my desk in my living room next to my couch. And you know, that actually made it so that I overworked [. . .] the impetus was always like, oh well, I can just stay online and keep working. There's nobody who's kind of like sitting next to me, going home. There's nobody saying like, Sophie [name changed], you're here awfully late. So, I actually overworked. And so, I moved to a two bedroom so that I could get separation from my office.

When it comes to regulating work, Sophie is skeptical of relying solely on self-discipline. She would rather have external controls in place to prevent "overworking," though, in the absence of such controls, she turns to personal strategies to create boundaries between work and life. Sophie's desire to establish a differentiation between work and life can be interpreted as a reaction to the pathological flipside of the entrepreneurial self. According to Ehrenberg (2010), the entrepreneurial ideals of flexibility and autonomy often led to a "weariness of the self." Burnout and depression amounted to major psychological and economic threats for white-collar workers (Neckel and Wagner 2013), especially in highly digitalized work settings (Christiaens 2020). In particular, tech workers perceive their mental resources to be at risk from the possibility of never really logging off. Aside from yoga and meditation (which are stereotypically common among my interviewees), tech workers try to avoid the weariness of the self by drawing boundaries between work and life.

Like most tech workers, Sophie does not turn to collective bodies to address the blurriness between work and life. If businesses do not take care of it, the lack of differentiation between work and life is not considered a problem that requires institutional or political actions. Rather, it demands individual responses. Here we can clearly see how certain neoliberal principles live on. Nevertheless, the individual responses are formulated within a frame that problematizes structural conditions. Mindfulness echoes elements of tech workers' critical worldviews through the more or less explicit problematization of societal pressures. In this sense, mindfulness would be mischaracterized as a genuinely neoliberal code of conduct (Leggett 2021). Currently, however, mindfulness only *challenges neoliberalism from within the epistemic doxa of self-control and self-responsibility*. The mindful schemas among tech workers do not achieve a genuine break with the individualistic knowledge orders that characterize entrepreneurial selves.

Such an ambivalence is also present in tech workers' relationship with social media. On the one hand, most tech workers are very active on social media. In particular, I find that LinkedIn has become a crucial social media network for tech workers to build connections, promote their work, and identify new job opportunities. Most tech companies also have an internal social media channel, such as Slack, to which tech workers also contribute content (often simply as part of their jobs). Furthermore, tech workers from abroad are typically dependent on social media for their personal relationships. They often spend even more time on social media due to long distance relationships with partners, family, and friends. A lot of tech workers are therefore ambivalent about social media in the sense that they value and rely on its services while simultaneously being unhappy about their amount of "screen time." It becomes clear that even those responsible for programming and designing our digital tools and virtual worlds are wary of spending too much time online. Many of my interviewees also problematize social media content, worrying about disinformation as well as a loss of "true" connection to the social world. While the organizational self feared the alienating metropolis (Simmel 2006), and the entrepreneurial self feared alienating corporations (Marcuse 2013), tech workers fear a loss of resonance with the world, other people, and themselves by getting detached from "real life." Again, though, this problematization of structural conditions does not lead to collective action but primarily to individual responses, even among radical tech workers. Even though tech workers are well positioned, for instance, to push for the design and programming of less addictive social media applications, this possibility does not come up as a topic of debate when discussing their worries about social media. Instead, one popular response is the temporary retreat from digital immersion—the so-called digital detox or practice of de-networking (Karppi et al. 2021). This response often takes place through spending time in "nature." Max, for instance, a UX designer from Germany, told me:

> I love to get away. I famously among my friends, I famously went to an island in the Pacific Ocean once, at which they don't even have running water, you know, just to get away from technology.

Tech workers aim to avoid not only economic exploitation and alienation but also an exhaustion of their mental energies. For Max, breaking with routine and stepping away from the constant immersion in digital life every now and then is essential. While this coping strategy is surely not exclusive to tech workers, it is important to take into account in understanding their mindset because it can be considered emblematic of their reflective but heavily individualistic conduct of life. Though my interviewees belong to the group that designs and programs the digital infrastructure of modern societies and possess the expertise to propose alternatives, they generally do not see the challenges of working and living in a heavily digitalized world as issues to be addressed collectively. While there are some gestures toward institutional engagement in their general subjectivation—evident, for instance, in occasional protests, open-source activities, practices of workplace solidarity, and emergent union formations—the typical tech worker subject I encountered remains firmly anchored in an individualist modus operandi.

Aside from this insight, Max's comments also suggest the importance of "nature" for tech workers. A number of my interviewees reported that they try to spend a lot of time in nature outside of work, whether through vacations or sport. This further challenges the stereotype of tech workers as isolated nerds confined to their offices or basements. On the contrary, like many other groups, tech workers are drawn to outdoor activities as a way of balancing their screen-focused working lives. In terms of grasping tech workers' ecological orientations, it is also important to be aware of their pronounced worries about climate change. When I asked the final question of my interview—"What is your gut feeling, are future generations going to have it harder or easier in life than your generation?"—a slight majority of my interlocutors were pessimistic, reporting *climate change* as their greatest concern. While one might expect the coders of our digital worlds to champion technological utopias—and indeed, about a quarter of my interviewees held such views—concerns about climate change were far more common. Moritz, a data scientist from Germany, provides such a response:

> I would say we're harming the planet at the moment. I personally hope that I'm still here for another sixty years. So, I want to have a good environment where I can live in. And obviously also, all generations younger than my generation will spend even more time here. And I like to say it in hard words: I don't see why we should destroy this planet and don't think about generations after us. I think that's just not fair.

Moritz voices his concerns about climate change, connecting them with moral principles. While he initially expresses worry about the state of the environment for his own sake, he is quick to add that his concerns also extend to future generations. He positions himself as guided by the principle of contributing to the welfare of others, which Lamont (1992) considers the basis of moral boundaries. Moritz's ecological mindfulness thus aligns with the wider moral underpinning of tech workers' boundary-makings. Furthermore, in the case of Moritz, eco-moral mindfulness translates into an intense inner relationship to an ecological consciousness: "Something which lets me feel bad [is that] I'm still not a full vegetarian. So, I like eating meat, I'm giving my best to reduce it as much as possible, but I'm still not fully vegetarian. So, I'm betraying myself there already a bit." Moritz critically interrogates his own contributions to climate change against the moral background of being mindful of the vulnerability of the natural environment. What seems most important to him is trying his best up to a certain point. Moritz cultivates an ideal of finding a balance between his lifestyle and his CO_2 emissions. This is also the case with Jennifer, the data scientist from San Francisco we met at the beginning of this chapter. She told me:

> Starting in undergrad, I was vegan for like four years. I'm no longer vegan any more. But that was one of the things I tried to do to you know, combat climate change. Now that I'm here, I mean, I chose to live where I live one because it was nice, because it's by the ocean, but two, it's a mile from the office, so I don't even have to drive. So, I just walk or bike to the office, which is really nice. I try to you know, use reasonable things. I cook at home. Little things here and there that I try to implement my life. I could be better than I am. But I guess I try the best with what I have for now.

Like many other tech workers, Moritz and Jennifer scan their lifestyle practices and evaluate them against criteria of the right ecological way of life. At times, it even seems that Moritz and Jennifer take up the role of an inspector or judge when reflecting on their tastes and leisure activities. Deviating from post-Fordist ideals of the self, this subjectivity indicates a new kind of ecological polis (Chiapello 2013) that manifests in forms of self-limitation. At the same time, this modus operandi may not be entirely unselfish in the sense that it can generate symbolic profits within certain fields (Neckel 2017).

Moreover, while tech workers present an honest longing to align their lifestyles with principles of care and limitation, they remain attached to the epistemic doxa of self-control and self-responsibility. This also holds true in their responses to climate change. Not even among very critical tech workers, who express sympathy for democratic socialism and could not imagine working for Big Tech companies, is there a widespread orientation toward institutionalized bodies of the civil and political sphere. While the mindful lifestyle clearly builds on critical orientations, it ultimately mobilizes individual rather than collectivist responses against structural relations of mental and ecological exploitation.

4.4 LIFESTYLES AND INSTITUTIONAL DISCOURSE

Now, let us turn our attention to consider how lifestyles are thematized in the institutional interpellation of tech workers. This section investigates the extent to which institutional discourses echo tech workers' self-presentations. It is important to acknowledge up front that the textual materials I draw on, namely study programs and job ads, are not much concerned with addressing lifestyles explicitly. However, as I will show, there exist a number of indirect and implicit valorizations of lifestyles in the discursive interpellation of tech workers. This underscores the sense that certain aspects of tech workers' social codes are being co-opted.

In the economic field, I find that tech workers' self-codification as comfortable explorers is echoed within institutional-discursive interpellations in a manner that aligns lifestyle aspirations with the economic operations of tech firms. The German start-up TaxFix, for instance, states in their job ad that the ideal UX designer should "push boundaries and explore new territory." Tech workers are imagined as fundamentally curious subjects who are capable of discovering new sources of revenue. Under the rubric titled "Why TaxFix," the job ad lists all the perks that designers would enjoy at their firm:

- A chance to do meaningful, people-centric work with an international team of passionate professionals.
- Holistic wellbeing with mental health coaching sessions, a discounted membership to Urban Sports Club, and supplemental child care support.

- Employee stock options for all employees because everyone deserves to benefit from the success they help to create.
- Full trust to take ownership of your work in a flat hierarchy where feedback is encouraged and expected.
- Dedicated relocation and visa support for those that need it.
- The freedom to work from home or our modern office plus healthy drinks and snacks when you do come in. (TaxFix, GER, UX).

The job ad extract illustrates the significant efforts made by tech companies to brand themselves as employers who provide comfort and support for their explorative employees. The level of comfort that is offered includes material and recreational aspects as well as the moral comfort of "meaningful" and "people-centric" work. This combination works to embed tech workers' self-understanding as comfortable explorers within a framework of productivity. Tech firms aim to demonstrate that they are attuned to tech workers' desire for comfortable exploration while subsuming these desires within their overarching economic performance goals.

Within the academic field, tech workers are also interpellated quite explicitly as comfortable explorers. For instance, the study program description for the master's in data science at the University of Virginia opens with the following interpellation:

> You don't settle for the first answer you receive. Or the simplest one. You can dig deeper than most people, and you enjoy puzzles and pondering the big questions. You relish the challenge of diving into a massive amount of information and surfacing with usable and actionable insights. (Virginia, US, DS)

Like many other universities, Virginia imagines data scientists as explorative subjects. While universities do note that challenges arise from the growing amount of data, they are just as keen to repeatedly emphasize the job opportunities that emerge from this: "To meet the growing demand for skillful, agile data scientists, the Master of Science in Data Science (MSDS) degree at the University of Virginia offers a rigorous 11-month professional master's program." The academic discourse constructs tech workers not only as discoverers of hidden patterns and latent needs but also as individuals equipped with the skills to carve out a comfortable position within a competitive labor market. Across the analyzed discourses of the economic and academic fields, there is a pronounced echo of tech workers' code of comfortable exploring.

The social code of ordinariness, on the other hand, is much less present in the discursive interpellation of tech workers. I found that neither established tech companies nor start-ups engage in framing tech workers in semantics that invoke ordinariness. Rather, the tech industry codifies its high-paid workers as *unique* professionals; the average tech worker is not imagined as an average person. This may be interpreted against the backdrop of tech companies framing their products and services as "disruptive" or "revolutionary." Codifying employees as "unique" rather than "ordinary" allows them to establish an elective affinity between business model and workforce. The German tech giant SAP, for instance, explicitly connects the innovative technologies they aim to create with the personality of their employees: "To harness the power of innovation, SAP invests in the development of its diverse employees. We aspire to leverage the qualities and appreciate the unique competencies that each person brings to the company." Like in other ads, extraordinariness rather than ordinariness is the dominant theme. The economic field establishes a connection between its powerful technologies and the (diverse) personhood of its employees.

Study programs are less concerned with constructing tech workers explicitly as special or extraordinary. Nevertheless, the universities also do not imagine tech workers as regular and down-to-earth employees who work from nine to five. Instead, as mentioned in the section on hybridity, universities prefer to classify tech workers as "leaders" who are driven by a restless will for innovation and disruption. The program description at the University of Virginia, for instance, features the lines, "Now more than ever, the field of data science demands leaders with diverse backgrounds." The master's in interdisciplinary data science at Duke is described as a "home for creative problem-solvers who want to use data strategically to advance society. We're cultivating a new type of *quantitative thought leader* who uses computational strategies to generate innovation and insights" (emphasis added). This codification scheme, working across institutional discourses in the academic and economic fields, is at odds with tech workers' dominant self-understandings, which valorize the idea of the regular employee with a stable work–life balance over that of the leader.

If at all, the social code of ordinariness surfaces latently in the discourse through promises that tech workers will be part of the moral mission of providing widely accessible technologies. A great number of tech companies and universities brand themselves as places where tech workers can (learn to) develop accessible digital technologies and solutions for society at large. The German start-up Healy, for instance, frames itself as developing products "for everywhere, at all times, and for everybody." In such discourses, tech workers are figured as a highly skilled workforce that remains in touch with the normal people whose needs they aim to address. Overall, however, extraordinariness rather than ordinariness is the dominant mode of interpellating tech workers within the economic and academic fields.

Finally, let us turn to the code of mindfulness. In the economic field, the code of mindfulness finds its way into the discourse of tech companies through so-called perks. We can make out an echo of tech workers' mindfulness in tech firms' listing of benefits such as massages and spaces for yoga as well as in their general branding as habitats that allow for healthy work–life balances. Working for a tech company is advertised as an experience that does not exhaust mental capacities. Interestingly, though, the listing of perks sits at odds with tech workers' preference for differentiating between work and life. At the German tech company Delivery Hero, for instance, mindfulness ultimately appears as a code that is mobilized to keep employees longer at work: "Enjoy massages, get your haircut in the office, join our free yoga classes or take a timeout in our nap room." Here, we can see how the tech industry aims to align the longing for mindfulness among tech workers with the business interests of the company. It seems that just as some tech companies want to establish platform ecosystems whose users will be disinclined not to leave (Staab 2019), so do a number of tech companies want to establish worlds of work where employees do not feel the need to leave the workplace. The perks offered thus serve not only as a means of attracting talent but also as a strategy for increasing employees' work output. From a Marxist perspective, the provision of perks can be understood as a tactic to control and maximize the transformation of labor power into actual labor.

As Wu (2024) points out, tech companies face the challenge that creative output from knowledge workers cannot be forcibly extracted; it must be nurtured and stimulated.

The academic field, on the other hand, does not thematize the issue of caring about psycho-mental resources in its study programs. It is only the issue of caring about ecological resources that is addressed in the analyzed materials. More specifically, a number of universities link tech work with climate action. The University of Potsdam, for instance, promotes its master's study program as opening pathways into "climate research." Studying data science or UX design is branded as a means of acquiring the professional tool kit to "tackle the world's biggest problems" (University of Virginia, master's in data science). Similar to the economic field, the academic field addresses climate change as an ecological issue that must be tackled at the interface between knowledge and technology.

Even though the discourse analysis is limited to study programs and job ads, it can be tentatively concluded that, again, there is only a partial correspondence between the self-understandings of tech workers and their discursive codifications by universities and tech companies. The lifestyles of tech workers do not fit neatly with how the academic and economic fields imagine their personhood. While the codes of comfortable exploring and mindfulness find pronounced echoes in the discourse, they are also subject to significant reinterpretations. In the economic field especially, the relationship between the self-understandings of tech workers and the institutional-discursive interpellation of their subjectivity indicates processes of appropriation and endogenization on the part of the firms. These processes manifest in the way that the economic field echoes the self-codifications of tech workers only to the extent that these can be aligned with logics of commodification.

5 CONTRADICTORY CLASS FORMATION

> It is in these intermediate zones of social space that the indeterminacy and the fuzziness of the relationships between practices and positions are the greatest, and that the room left open for symbolic strategies designed to jam this relationship is the largest.
>
> —Bourdieu (1987, 12)

The previous chapters have used interview data to reconstruct the central social codes of tech workers, those professionals who program our digital worlds and occupy the intermediate zone in between tech entrepreneurs, on the one side, and highly precarious digital workers on the other. Building on this foundation of empirical analysis, we will now bring the theoretical question of *class formation* into focus. Anchored in the cultural tradition of class analysis, this chapter explores whether the social codes of tech workers translate into the making of a "realized class". Are tech workers more than a "class on paper"? Can they be understood as a distinct social formation in the class matrix?

A class on paper, also referred to as "probable group" or "statistical group," refers to a collection of individuals who share similar positions in terms of economic and (institutionalized) cultural and social capital (Bourdieu 1985, 1998). Through this lens, objective class positions are determined with reference to the individual's location within a multidimensional space of resources. Tech workers can be considered to compose a class on paper in the sense that, despite intra- and interprofessional differences, members of this segment typically occupy a delimited space within the class matrix.

They have similar earnings, academic credentials, forms of expertise, work characteristics, geographic locations, and access to networks. Located within the intermediate zone of the class matrix, tech workers share a common space of possibilities that defines their life chances.

What, on the other hand, are the characteristics of a realized class? Bourdieu's early work emphasizes the role of lifestyles in the process of class formation. In his landmark study, *Distinction* (1984), he demonstrates how shared everyday tastes and activities allow for the making of distinctive sociocultural classes. Later, he broadened his conceptualization of "real" or "realized" classes toward processes of *classification* more generally (Bourdieu 1998). This emphasis on classification has become central to the evolving cultural tradition of class analysis in recent decades (Lamont 1992; Reckwitz 2020; Skeggs 2004; Westheuser 2020). While still influenced by Marx's distinction between a class "in-itself" and "for-itself" (Andrew 1983; Marx 1963), the field has largely abandoned the often desperate search for a revolutionary subject and put aside teleological notions of consciousness. Instead, the focus has shifted to the ways in which contingent symbolic boundaries—formed on the basis of classification systems—shape and maintain social divisions. Importantly, though, contemporary cultural class analysis continues to maintain that the full realization of a class requires collective action and struggle; there is no complete break with the Marxist tradition in this sense. The key difference is that, in opposition to Marx, the cultural sphere takes center stage, conceptualized as the engine room of class society—the arena where class identities are forged and where collective struggle is either enabled or hindered. Inspired by Weber (1968), cultural class analysis examines how conducts of life—understood as values, roles, and lifestyles—shape the stratification of society.

Viewed through this lens, the existence of a class on paper is fundamentally a question of resources, while the formation of an active or realized class hinges on the cultural and symbolic realm that is connected to, but not simply determined by, objective positions. Class—which, together with the interrelated dimensions of gender and race, constitutes the primary matrix of social inequality (Hall 2019; Skeggs 2004)—is defined by a dynamic interplay of positions and positionings. In the end, class is always relational and processual, arising from the relationship between objective resource-dependent positions, on the one hand, and subjective codifications on the other (I use the term codification rather than classification to highlight the normative dimension of this process).

What codification principles, then, are required for a cluster of individuals to come together as a realized class? My analysis builds on the cultural tradition to suggest that the process of class formation necessitates the sharing not of any social codes but rather ones linked to three key principles: (1) *reflection*, (2) *collectivity*, and (3) *solidarity*. That is to say, the process of class formation requires: first, knowledge and awareness of power relations and exploitation; second, a significant level of affiliation; and third, the active demonstration of support for a social group (either one's own or another). These three principles can be taken from cultural class scholarship as the key mechanisms through which a statistical grouping is able to metamorphose into an active and relatively autonomous social force.[1] Any number of individuals with a relative degree of commonality in terms of economic, cultural, and social capital can constitute a realized class if they cultivate common reflections, forms of collectivity, and manifestations of solidarity.[2] Of course, the greater the disparity between their economic, cultural, and social positions, the more symbolic labor is needed to forge a unified class out of a number of individuals (Bourdieu 1985; Lamont 1992).

This chapter—and indeed the rest of the book—shifts from presenting new empirical material to systematically theorizing the social codes reconstructed in earlier chapters. This will allow for a more elaborate and contextual understanding of tech workers' class formation, one that I argue is profoundly contradictory: On the one hand, tech workers form a distinct social group that is surprisingly cohesive, culturally, morally, and politically speaking. In the US and Germany, across the professions of data science and UX design, principles of reflection, affiliation, and solidarity are commonly cultivated and shared in ways that establish boundaries vis-à-vis other social classes. Clearly, then, tech workers are more than a class on paper. Yet, on the other hand, tech workers remain attached to individualist modes of perception and recognition that undermine collectivity. There is a *disconnect between their social critiques and the actions they report, and gender and racial divisions further limit the potential for greater unity.* In addition, some of their workerist class codes, intended to express solidarity with lower classes, paradoxically hinder upward social mobility. This is not to go so far as to say that tech workers display ironic forms of class consciousness or self-awareness (Chouliaraki 2013; Savage 2015, 134), and it is important to recognize that many of these internal contradictions are not entirely unique to tech workers. In this way, coming to terms with tech workers' class formation provides us with valuable insights into middle-class dynamics more generally.

Accounting sociologically for tech workers' contradictory class formation, I argue, requires examining their experiences of *multiple crises* as well as their *specific socio-structural backgrounds and positions.* Tech workers' class formation must be contextualized against the backdrop of three central crises: (1) the economic crisis of neoliberalism, (2) the political crisis over legitimate representation, and (3) the ecological crisis of climate change. Experiences from these crises run from the 2007–2008 financial crash, whose aftershocks were felt by many tech workers when they entered the job market; the political upheaval triggered by Cambridge Analytica and Trump's election in 2016; and to deep-seated concerns over climate change. The third section of this chapter discusses how tech workers' socio-structural positions and backgrounds (especially in terms of class, gender, and race) inform their subjective positionings within these crises. This dual perspective on situations and resources will grant us a thick understanding of how and why market- and creativity-oriented ideal-typical selves are undergoing a convoluted transformation. This chapter concludes with an analysis of *interprofessional and national differences.* While the previous chapters have focused on the overwhelming similarities between data scientists and UX designers across the US and Germany, the final section of this chapter systemizes key differences and discusses how these contribute to the contradictory class formation of tech workers.

5.1 DOUBLE-EDGED CODES

The subjectivity of tech workers presents us with transfigurations of the so-called entrepreneurial self, which has been understood as the ideal-typical subjectivity of middle-class professionals under post-Fordism (Bröckling 2015; Pongratz and Voss 2003; Rose 1989; Skeggs 2004). Tech workers do not only deviate from neoliberal cultures and entrepreneurial ideals—they actively construct boundaries against them. The notion of a subject whose normative capacities are exhausted in a quest for economic wealth and creative self-fulfillment does not provide a comprehensive account of the hearts and minds of tech workers. While they remain attached to forms of self-control, flexibility, and autonomy, tech workers cultivate a number of new social codes within the domains of worldviews, professional roles, and lifestyles. Across these spheres, a distinct social character is contoured through tech workers' quest to be middle-class wealthy and morally worthy.

Our concern now, however, is systematically to interpret this identity in terms of class. Do the social codes of tech workers build on reflection, collectivity, and solidarity—the three pillars of a realized social class? And what specific group-based boundaries are summoned by tech workers' symbolic boundaries? In response to these questions, let us first consider the return of social critique. This worldview signals that tech workers spend a significant amount of time reflecting on issues of power and domination. They perceive society not as a flat network but rather as a hierarchical system. Notably, it is those tech workers who lean toward more radical and reformist ideal types—who together account for roughly three-quarters of my sample—who actively contemplate their specific position within systems of domination.[3] They do not consider their interests to be aligned with those of the owners of tech firms. This sense of distance from capital also manifests in my interviewees' widespread self-classification as "workers"—pointing to a group-based boundary vis-à-vis tech bosses that indicates both a reflection of power and an articulation of affiliation. Furthermore, the social critique is manifested in boundaries drawn against other upper-middle-class professionals, such as bankers or consultants, who are deemed less concerned with doing socially meaningful work. Tech workers often morally devalorize other professionals who seemingly do not seek to align their careers with other-oriented values. This suggests that tech workers constitute a distinct upper-middle-class fraction shaped through moral boundaries, even if, as I will later explore in detail, some of their moral codes are not entirely unique but rather reflect wider value shifts in light of societal crises.

There is, however, a flip side to the role of social critique in this class formation. The boundaries drawn through social critique often create a paradox, as they remain rooted in individualistic principles and are primarily expressed through symbolic displays of the self. Tech workers rarely report translating their moral beliefs into corresponding actions. While it seems to be part of the human condition that the dimensions of self-understanding and practice rarely correspond fully, there is something striking about the extent to which tech workers' pronounced moral codes appear to remain aspirational. The gap between their expressed values and reported practices suggests that while tech workers identify with ideals of social justice and critiques of power structures, there is very little evidence of these commitments being enacted in tangible, impactful ways. This discrepancy highlights a fundamental tension whereby tech workers articulate a clear rejection of the status quo and align themselves with worker identities, yet often struggle to transform this ideological stance into meaningful action that challenges existing power structures. For example, tech workers rarely report engaging in practices such as forming unions or participating in protests. While three-quarters of my interviewees convincingly told me they could imagine joining a union, only two could say they were currently members. Other, less formal kinds of collective action were likewise seldom mentioned. In other words, tech workers appear to exist primarily as a symbolic community and not as a community in action. They tend to ask themselves what individual, rather than collective, actions they can undertake to change things. This individualist and deeply incorporated modus operandi can be considered largely responsible for the fact that recent layoffs in the tech industry, as well as increasing alliances between tech leadership and the second Trump administration, are not met with more resistance by tech workers. Furthermore, due to their pronounced status aspirations, tech workers are highly concerned about the risks involved in activism and the perceived low odds of being able to make a difference. For most tech workers, it is inconceivable to leave behind the economic status of the middle class (a comfort the majority of them have grown up with). To avoid risking this comfort, they are willing to make significant sacrifices with respect to their moral praxis. Tech corporations are aware of this. As Rothstein (2022) has argued, tech firms put a lot of effort into convincing their employees that unionization efforts would undermine competitiveness within markets. In the case of tech workers with equity (which is especially the case at start-ups), additional incentives encourage workers to side with the interests of capital (Neff 2012).

This is not to say that there aren't any shared practices among tech workers through which they attempt to translate their moral outlooks into action. This becomes evident when we consider their voting behavior, which is coherently liberal/center-left in orientation. An overwhelming majority of my interviewees told me they vote for the Democratic Party or center-left parties (especially the Green Party) in the US and Germany, respectively. We cannot discount the possibility of interviewer effect, here—that is, interviewees may have been less likely to voice more conservative values they might assume a young sociologist would not share. Nonetheless, my empirical observations here are reinforced by a quantitative analysis of tech workers' contributions to political campaigns, which presented similar findings (see Selling and Strimling 2023). Furthermore, some tech workers reported engaging in practices of workplace solidarity such as establishing pay transparency as well as in engaging in open-source activities. However, reports of collective action remain the exception rather than the rule, suggesting that the transformative potential of tech workers' codes has yet to be fully realized or mobilized on a larger scale. This said, it is important to reckon with the predominantly symbolic manifestations of tech workers' social critique. From a social-theoretical standpoint, therefore, I suggest conceptualizing tech workers' cultivation of social critique as one key element within their contradictory class formation.

Another instance in which the subjectivity of tech workers indicates the formation of a contradictory class fraction is through the critique of insufficient diversity. As I have outlined, tech workers frequently problematize a lack of diversity and equal representation in the world of tech and in society at large. They speak of tech companies as not doing enough for gender and racial equality. My interviewees generally articulated a strong sense of solidarity with marginalized social groups. They rarely cultivate blatant "post-racial rationalizations", understood as frameworks where race is not regarded as a societal issue but replaced by a belief in an already existing colorblind meritocracy (Meghji and Saini 2018). Furthermore, most tech workers expressed concerns about the inscription of bias into digital technologies (see also Brown et al. 2024). They signal awareness of their inscription power and embrace a relatively reflective relationship to the digital means of production. Thus, while tech workers are not techno-pessimist but sustain a general belief in the progressive potential of digital technology, it is important to highlight that they do not cultivate a naïve "technological solutionism" that is emblematic of the entrepreneurial class and their "Californian Ideology" (Barbrook and Cameron 1995; Burrell and Fourcade 2021). Some of my interviewees even call for more government regulation to tackle issues of data bias.

At the same time, however, it is important to emphasize that discrimination still exists in the field of tech work. Women, People of Color, and individuals from non-academic family backgrounds are significantly underrepresented. Moreover, my research revealed symbolic mechanisms through which this underrepresentation is reproduced, including a subtle form of "techno-masculinity" (Poster 2013). This techno-masculinity enfolds power through symbolic hierarchies between professions. As outlined, the profession of UX design, which employs a higher proportion of women, is often codified as less "technical" and more "creative" vis-à-vis the profession of data science. This codification can be said to contribute to UX designers' lower average salaries and their heightened sense of a need to justify the value of their expertise. So, even though tech workers attach great symbolic value to the principle of inclusion, there also exist social codes that produce inner divides through exerting subtle discrimination. This copresence of codifications that valorize diversity alongside those generating latent forms of discrimination contributes to the contradictory class formation of tech workers and may undermine, or at least frustrate, the further consolidation of their fragile collectivity.

Hybridity is the defining logic at play in the professional roles of tech workers. One manifestation of hybrid professionalism lies in tech workers' combination of the generally conflictive character slates of the nerd and the communicator. Tech workers bond with each other by valuing a combination of technical and communicative capacities in one another. In terms of class formation, I argue that this figuration allows tech workers to affiliate as a distinct grouping in the competitive "system of professions" (Abbott 1988). By cultivating post-nerdiness, tech workers are able to draw boundaries vis-à-vis seemingly more one-dimensional professionals, such as statisticians or traditional computer scientists (Kendall 1999). Interestingly, tech workers do not link this skill set with any specific educational background. They establish low formalized entry boundaries to their own guild. It seems that more informal codes are, however, mobilized to establish occupational closure. Here again we see a contradictory logic of a class formation, which can be said to operate and enfold power via a paradox mode of inclusive domination in this case. Notably, inclusive domination assumes another form for tech workers from the Global South, who must deal with differential accessibility to tech companies (and promotions) due to the constraints of visa regimes. Furthermore, cultivating the identity of the post-nerd is more difficult for tech workers from the Global South, who are effectively classified as less skilled in communication. Across such instances, we can see how "race is [. . .] the modality in which class is 'lived'" (Hall 2019, 216). Put simply, race functions as a structural and ideological mechanism that divides members of one occupational segment (see also Amrute 2016). Finally, an overarching manifestation of hybridity is instantiated in tech workers' approach to balancing distinct needs. My interviewees seek to align their worldviews with their organizational embeddedness through brokering users' needs and those of business. This particular logic of hybridity links to class formation in the sense that it affiliates most tech workers in an orientation toward class compromise. Their aspiration to be both middle-class wealthy and morally worthy culminates in a self with *disciplined dissonance.*

In terms of lifestyle, the leisure and consumption activities of tech workers tend to be structured by codes of ordinariness, mindfulness, and comfortable exploration. The emphasis on an ordinary lifestyle presents a particularly contradictory aspect of tech workers' class formation. In one sense, the code of ordinariness builds bridges toward less affluent classes. Tech workers' foregrounding of down-to-earth leisure activities such as walking the dog or spending time with the family signals inclusiveness, and thus holds the potential for affiliation with "workers" more generally (Jarness and Flemmen 2017). This distinguishes tech workers from those professionals within the middle class, who are particularly present in the creative and cultural industries, who primarily strive for singularity and extraordinariness (Reckwitz 2020). In another sense, though, tech workers still cultivate a set of exclusive lifestyles, such as cosmopolitanism and a preference for pricy urban areas—habitats, in which they then display their down-to-earthness. Moreover, the foregrounding of a lifestyle of ordinariness may function as a "strategy without strategists" (Foucault 1978, 132) for securing symbolic capital (Jarness and Flemmen 2017). Drawing on the work of Friedman and Reeves (2020), who diagnose this strategy among British elites, but extending their analysis to the upper-middle-class, I argue that by valorizing cultural inclusiveness, tech workers avoid appearing elitist, thus shielding themselves from critique at the same time as they rise above a generally declining middle class in economic terms.

The lifestyles of comfortable exploring and mindfulness also propel the formation of tech workers as a contradictory class fraction. The lifestyle of comfortable exploring indicates that tech workers want both economic security and creative self-actualization. This results in a contradictory figuration: Tech workers refuse the Faustian pact of post-Fordism (in which demands for more flexibility and creativity were partly achieved at the cost of economic returns on labor) while remaining attached to a set of individualistic values that draw them away from the very institutions and organizations (such as unions) that would be necessary to genuinely actualize their demands for economic security. In a similar vein, the ethic of mindfulness also indicates the formation of a contradictory social class. At first sight, mindfulness signals a reflective ethic through which tech workers seek to avoid burnouts, fatigue, or other disruptions to the functioning of the self. Through this technology of the self, tech workers aspire to take care of their mental resources by instantiating a redifferentiation between work and life. However, it also presents a highly individual approach to tackling the pathological effects of the current socioeconomic structures. This individualized approach is also the case with tech workers' lifestyle responses to climate change.

In this examination of worldviews, professional roles, and lifestyles, *The Social Codes of Tech Workers* demonstrates how tech workers form a contradictory class fraction that is only partially realized. Tech workers cultivate class codes that are double-edged in the sense of operating in different and often conflictive ways. On the one hand, we have seen that the social codes indeed relate (to different degrees and in different ways) to the principles of reflection of power, affiliation, and solidarity. In other words, tech workers are not departing from the entrepreneurial self or the related "Californian Ideology" (Barbrook and Cameron 1995) in any sort of fashion. The transformation of the entrepreneurial self is undergirded by principles of reflection, collectivity, and solidarity. On the other hand, I also showed how the transformative social codes go hand in hand with a neglect of certain power relations, a tendency for individualism and symbolic claims, and the latent pursuit of self-interest. That is to say, tech workers are more than a statistical grouping of similarly positioned agents, but less than a collectively orchestrated ensemble of actors in pursuit of shared interests.

5.2 SITUATIONS OF CRISIS

Let us now systematically turn to the question of how this contradictory class formation can be accounted for. How is it that tech workers—who constitute a rising professional segment in the contemporary economy—are questioning and partially departing from the ideal of a market-oriented self?

In the previous part of the book, I have presented scattered clues as to solving this puzzle. I presented various interview extracts where my interviewees discuss how different crises starkly influenced their lives and careers. Here, I argue that we must focus our attention on the *multiple situations of crisis* with which tech workers have been, and still are, confronted, in order to explain their constituting a contradictory class fraction. Crises are a type of situation that can challenge even the most sedimented social orders and codes. Crises are objective insofar as they set external limits to our lines of action or plans (Beck 1992). Nevertheless, drawing on the pragmatist tradition, a crisis is not a fact we come to know but something that we need to make sense of (Raza 2022). While crises are surely not the sole force that accounts for social change, the event of a crisis must be considered a specific context that enables or accelerates the transformation of even the most unquestioned forms of life; a crisis puts the common sense of a conjuncture up to debate and may thus allow for the detection of deeply habitualized forms of subjectivity. With regard to tech workers, their distinctive social codes and contradictory class formation must be considered against the backdrop of three crises: (1) the economic crisis of neoliberalism, (2) the political crisis over legitimate representation, and (3) the ecological crisis of climate change. All three of these crises have sparked disillusionment among tech workers with the current organization of economy and society.

The *economic crisis of neoliberalism* undergirds the contradictory class formation of tech workers in several ways. First, significant parts of the contemporary generation of tech workers are scarred by their personal experience of the financial crisis of 2007/08. As reconstructed in the preceding empirical chapters, the last major economic crash hit many tech workers at a decisive moment in their life. As the mean age of my interviewees is thirty-four, a large number of them entered the labor market just after the financial field catapulted the global economy into a recession. This made entry into the labor market very difficult for many of them and left a permanent mark on their outlook, which is to say, a distrust toward the market and parts of the capitalist elite. Given the global economic tensions at the time of writing this part on class, it will be interesting to observe how tech workers react to the restructuring of their sector. Sidney Rothstein (2022) has shown how in the past, tech workers have failed to organize when the discourse of management has succeeded in attributing layoffs to external factors. For Rothstein, the class struggles of tech workers depend on their capacity to recode discourses of management at work and as it pertains to work relations. My research indicates that the current recoding of management discourses among tech workers is insufficient. While tech workers clearly cultivate a more critical understanding of economic relations than did their forebears at the time of the dot-com crisis (Neff 2012), their critical talk carries too much individualistic baggage. Furthermore, it is important to recognize tech workers' strong attachment to a middle-class way of life. Notably, *this attachment is not only emotional but also deeply embedded in their material circumstances.* Many of my interviewees are economically tied to long-term financial commitments, such as mortgages, and have established lives in areas where rising living costs compel them to remain employed in the tech sector. In general, tech workers are therefore not willing to take their social critiques to the next level unless they are confident in their chances of success. It appears that there is a dual character to tech workers' experience of economic crisis: On the one hand, it propels critique, while, on the other hand, it creates a sense of appreciation for one's relative security within a decaying economic system.

What is clear is that tech workers experience capitalism as dysfunctional beyond the duration or aftermath of economic crashes: They believe the system is structurally flawed. Economic inequality, which has grown over the last decades in economies of the Global North (Milanovic 2016; Piketty 2014), is met especially with critique by tech workers. Tech workers typically do not consider justifiable the wealth of super-rich members of their sector such as Jeff Bezos. Additionally, they are critical of contemporary capitalism's tendency to blur the boundaries between work and life, which they link to burnouts. Furthermore, the proprietary market structures of the tech industry very much nurture the social critique of tech workers. The monopolizing tendencies in the field of tech, which have been cast as feudalist, have propelled a disillusionment with the neoliberal faith in markets. While tech workers are no revolutionaries or anti-capitalists (indeed, the critique of proprietary markets can be considered a genuinely capitalist stance), they are increasingly skeptical of deregulated markets. This is also the case with the housing market. Tech workers live in some of the most polarized housing economies in the world, such as San Francisco or Berlin. On a daily basis, they witness unprecedented levels of homelessness in overheated housing markets. The skyrocketing rents and housing prices not only disrupt beliefs in neoliberalized markets but also create a sense of a shared experience among tech workers, who relate to each other over increasing troubles to buy property despite earning upper-middle-class incomes. It is important to keep in mind, however, that some tech workers, especially the more senior ones, own property, pointing to an important economic and age-based internal differentiation within the occupational segment.

The second crisis that accounts for tech workers' contradictory class formation is the *political crisis over legitimate representation*. In recent decades, the systemic disadvantages of People of Color and women have garnered more and more attention in postindustrial societies, leading to intensive discursive struggles over racist and sexist structures (Brenner and Fraser 2017; Fraser 2009). As demonstrated in the previous chapters, tech workers echo arguments commonly put forward in these discourses, in some cases connected to direct experiences of discrimination and subjugation. It is also important to note that almost all of them have undergone university education and prefer to live in urban areas—two central factors that have long been identified as increasing one's likelihood of incorporating liberal values (Florida 2002). Furthermore, given tech workers' inscription power—which results from their influences on the design and functionality of digital technologies (Benjamin 2019)—they find themselves in a position where they must increasingly justify the way that algorithms differently shape the life chances of different social groups.

A pivotal political situation that propelled tech workers' turn toward moral worth, which is connected to their inscription power, is the political polarization that followed Donald Trump's election to the presidency in 2016. Tech workers underwent a class-specific experience of this event. This is because in late 2016, it became clear not only that the tech firm Cambridge Analytica had assisted Donald Trump's presidential victory through dubious data analytics procedures but also that Facebook's lax data policies and business interests had allowed the firm to easily mobilize personal data from Facebook (Venturini and Rogers 2019).[4] Suddenly, the belief that platform companies should disrupt established institutions lost its (remaining) appeal. This sparked a first wave of distancing among rank-and-file workers vis-à-vis the tech firms employing them. Then, in the aftermath of the 2016 election, most bosses of tech firms, who had predominantly supported Hillary Clinton's campaign, seemed not at all hesitant to build bridges with the newly elected Trump. Among my interviewees, these two instances inspired a sense of political and moral distance from tech management and the entrepreneurial class. The assumption that the tech industry was either progressive or politically neutral had been disrupted. Furthermore, the case of Cambridge

Analytica affected tech workers by confronting them with negative public perceptions of their work. In my interviews, I was told that after 2016, when one introduced oneself as "working in tech," people reacted differently than they had before. With the tech industry considered an aid to Donald Trump's election, the glamour and hopes around tech had turned into animus within the social fields that tech workers tend to mingle with (such as the academic field). As a result, the 2016 US election has become a key reference point in tech workers' narratives about their identity and aspirations. It marks a critical moment that distinguishes them as a class fraction (it remains to be seen what the long-term consequences of Trump's reelection as president in 2024, and the even closer relationships tech leaders are forging with his administration, will be for the class formation of tech workers).

I consider both the financial crisis in 2008 and the 2016 US election as *events* in the lives of tech workers in the sense given to this term by Badiou (2007). According to Badiou, an event takes place when an excluded part of a social order appears at the subjective level. By bringing to light new representations, an event provides the space for new subjects to emerge (Badiou 2007, 201). The event of the financial crisis, which occurred just as many now-mid-level tech workers were entering the labor market from the relatively safe haven of the university, fractured the neoliberal belief in the capability of markets to regulate themselves. The Cambridge Analytica scandal and the outcome of the 2016 US election can be considered an event in the sense that it disrupted a belief that various institutions of society, such as traditional politics and media, need to be replaced by new technologies—a belief that was virulent among tech workers before (Marwick 2013; Turner 2008). This ideological disruption has acted as a key driver behind the emerging labor activism within the tech industry. As Tan et al. (2023) have shown, reports of collective action among tech workers grew exponentially between 2017 and 2019.

However, the disruption unleashed by these events was clearly not so strong as to lead to a massive resignation from the industry or opposition to certain firms. Only roughly a quarter of my interviewees (those leaning toward the radical ideal type) told me they could not imagine working for the very big tech companies. Moreover, since 2019, reports of collective action have stagnated or declined (Tan et al. 2023). It must also be underscored that the political crisis has not led to a complete loss of techno-solutionist belief. There still exists the belief that, under different circumstances, digital technologies can significantly move society in a progressive direction. Moreover, the crisis over instances such as Cambridge Analytica has actually increased the *potential* of technosolutionist beliefs in a paradoxical way. As the narrative has emerged that digital technologies shaped the outcome of a US presidential election, the question of what to do with or about the tech industry has increased in societal relevance. The problematization of tech as having too much power or misusing their powers reproduces brandings by tech corporations that their technologies can actually disrupt the world. This promises to attract people with and without the ambition to change things from within the industry. The recent stagnation and decline of activism among tech workers may thus also be linked to a growing number of people entering the profession who are motivated exclusively by career prospects.

The third key crisis that must be considered to account for tech workers' cultural and moral orientations is that of *climate change*. As I have outlined based on interview extracts previously, tech workers are deeply worried about the planetary consequences of CO_2 emissions. When I asked the final open question of the interview ("What is your gut feeling, are future generations going to have it harder or easier in life than your generation?"), climate change was the biggest concern. The future imaginaries of tech workers are shaped by distinctly bleak ecological prognoses. Notably, however, the climate crisis does not impact workers' transformative social codes according to Badiou's logic of the event in the sense of a sudden rupture. Rather, the climate crisis shapes tech workers' social codes in a latent and long-term processual way. Further, the ecological crisis influences tech workers' trust in "free" markets. Given the prognoses by current climate models (which they, as academically trained and knowledge-oriented subjects, do not doubt), tech workers do not believe that deregulated markets in their current state will solve the challenge of global warming. At the same time, however, it is important to note that most tech workers do believe that digital technologies in the context of capitalist development hold potential to significantly reduce CO_2 emissions. Even though capitalism is often viewed as the source of these crises, belief in its capacity to generate solutions has not been genuinely eroded. *The crises tech workers are confronted with generate a certain disillusionment with capitalism, but this takes hold within a subjective framework that is itself disillusioned with imagining other modes of economic organization.*

Besides the economic, political, and ecological crises that have affected tech workers in distinctive and multifaceted ways, other crises surfaced in tech workers' narratives that contributed to their pronounced reflections on power relations as well as their (symbolic) longing for collectivity and solidarity. The Covid-19 pandemic, for instance, likely also contributed to tech workers' moral reorientations through widespread appeals to principles of solidarity at the time I conducted my interviews. However, I found that economic issues related to neoliberalism, the political crisis of legitimate representation, and the looming ecological crisis were the primary situational contexts that undergirded the new forms of subjectivity among tech workers. In the next step, I will consider how the socio-structural positions and classed, gendered, and racialized backgrounds of tech workers link with these situational contexts. The social theoretical endeavor behind this twofold analysis of the class formation of tech workers is to fuse perspectives from pragmatism, and notably its focus on *situations* (Abbott 2016; Boltanski 2011; Raza 2022), with a critical sociology concentrated on *positions* and underlying structures of domination (Bourdieu 1984; Friedman and Reeves 2020; Savage 2015).

5.3 SOCIO-STRUCTURAL POSITIONS AND BACKGROUNDS

The economic, political, and ecological crises that characterize contemporary social space produce situations in which tech workers reorient their subjectivity, in complex ways, toward principles of reflection, collectivity, and solidarity. However, situations and contexts are interwoven with the positions and backgrounds of actors shaped through multiple forms of capital. Thus, while crises tend to impact the entirety of a social space, actors experience the same crisis differentially depending on where they come from and what kind of resources they had and have at their disposal. Tech workers' contradictory transformation of the entrepreneurial self and their quest for moral worth must thus also be considered in the light of current and past life chances.

I have argued that tech workers experienced the financial crisis at a crucial age in their lives and are now embedded within an industry where the contradictions of capitalist production are quite pronounced. But we cannot overlook the privileged labor position that tech workers typically occupy in the current economic order. Given the general high demand for their labor power, most tech workers feel relatively secure on the labor market. Among my interlocutors, interviewed between 2020 and 2022, there was no widespread fear over being replaced by other humans or new machines in the near future. The specter of "deskilling" (Braverman 1974) does not (yet) haunt their minds. This lack of fear can be made sense of in the light of reports that 79 percent of tech workers who lost their jobs during layoffs in 2022 were able to find a new job within three months (Cambon and Guilford 2022). To be sure, a significant number of tech workers endure forms of short-contracting and most continuously invest time to stay up to date with the latest technological developments. Typically, however, their expertise is less threatened than is the case in other professional fields (Frey and Osborne 2017). The advent of ChatGPT and new Generative AI technologies may change this. What my study reveals is that between 2020 and 2022, there was a widespread sense among mid-career tech workers that another job is waiting around the corner—as long as one does not overdo it with challenging the bosses. I suggest that this relative lack of fear is a form of cultural and economic capital and is important to consider when accounting for the return of social critique among tech workers. The high demand for their knowledge expertise provides them with confidence and a safety net, allowing them to engage in certain forms of critique as well as symbolic practices of affiliation and solidarity. The class formation of tech workers suggests that within the middle layers of society, *it is security, rather than precarity, that fuels critique.*

Aside from their privileged economic position and the demand for their expertise, the class backgrounds of tech workers allow us to understand why they maneuver the economic crisis the way they do. A large majority of the tech workers I interviewed appear to have *middle- or even upper-middle-class backgrounds*, with parents pursuing high-status occupations such as law, medicine, or engineering. This includes many of the tech workers form the Global South that I interviewed. And while my interviewees have chosen a new social field in which to pursue their careers, there remains a deep aspiration to stay within the (upper) middle class. The social code of balancing different needs and the quest to live an existence that allows for middle-class wealth and moral worth must be set against this backdrop. While tech workers consider the current economic system to conflict with their moral values in manifold ways, they are not willing to step outside of the system. They are attached to a middle-class way of life and are willing to pay a certain price for it. The deep-rooted attachment to middle-class existence thus puts limits on tech workers' moral action, propelling the crystallization of a reformist class identity.

In some instances, though, we can identify positionings that hold the potential to exceed reformism. This is especially the case with the economic crisis around housing, where the class backgrounds of tech workers frame the experience of the housing crisis as an attack not only on their comfort but also on their class heritage. In San Francisco but also, increasingly, in Berlin, it is difficult even for well-paid tech workers to afford a family house or apartment. With most tech workers coming from middle-class backgrounds, there is a fundamental disillusionment with capitalism in the sense that, despite their high performances in the labor market, it is hard to maintain the living standard they enjoyed as children. In other words, there is a feeling of a *breach* of the capitalist bargain with the American Dream increasingly fading away for those who grew up with it. As mentioned, however, more senior tech workers are less affected by these processes as they are likelier to possess property purchased at significantly lower prices than those available now. Furthermore, the (mostly young) tech workers without property are not actually pushed out of the urban habitats they enjoy so much. Via the rental market, most are still integrated in the metropolitan networks that enable the quick and intensive accumulation of both cultural and social capital. These two aspects, likely, prevent a class formation that transitions from heterodoxy to heresy.

In addition to ambivalences, a critical study of subjectivation processes and class must also consider latent strategies. From this perspective, the critique of economic relations among tech workers is not only the result of a more-or-less rational reaction to economic crises, secure labor market positions, or middle-class backgrounds. Additionally, there is a latent legitimization strategy at play. Given that tech workers are increasingly pulling away economically within the middle layers of society, constituting one of the few middle-class fractions in the Global North that has thrived in the last twenty years, their critique of economic inequalities must also be interpreted as a strategy (without strategists) for avoiding the critique of others. By understanding and presenting themselves as critics of capitalism and as also enduring some hardships of the (housing) market, tech workers successfully sidestep the label of elites. The same applies to the lifestyle of ordinariness, which allows tech workers to avoid the image of the cultural snob in times of precarity. Zooming out, it becomes clear that there is a *double movement* to the class formation of tech workers: On the one hand, their social codes are linked to critiques of capitalism and the unequal distribution of different forms of capital. This can be considered an open boundary-making vis-à-vis the entrepreneurial class. On the other hand, the very same codes also allow them latently to shield and secure the relatively high returns on their labor power as well as their privileged class position. This can be considered a latent boundary-making vis-à-vis the economically lower social classes.

We can also witness the influence of capital and socioeconomic backgrounds on tech workers' navigation of the crises of legitimate political representation. As we have seen, tech workers are not only critical of economic relations but also of insufficient diversity in the tech industry and beyond.[5] While critiques were especially prominent among tech workers from marginalized socio-structural backgrounds, the problematization of a lack of diversity also regularly occurred among tech workers who have not experienced social marginalization across class, gender, or race. Generalizing tech workers as "tech bros" therefore appears questionable. I hold that the diversity codex among tech workers must be interpreted, from a sociological standpoint, as structured by their high volumes of cultural capital and their socio-spatial networks and habitats. All but one of my fifty-two interviewees hold an academic degree, and they have often spent multiple years in international and left-leaning areas of "creative cities" (Florida 2002, 2010). Tech workers can be categorized as cosmopolitans. My interviewees tend to have been very mobile during their education and their careers and can generally be considered to have profited from globalization (to different degrees, of course). Importantly, though, being spatially mobile is a capacity that comes with a higher cost for women than for men. Men are (still) more often in relationships with partners that are not working or working part-time and are therefore less geographically bound (Klammer 2023). This provides an important backdrop for understanding recent reports that show that male tech workers were 22 percent more successful in finding a new job after layoffs than female tech workers (Yosifova 2023).

Nevertheless, it is important to highlight that both male and female tech workers are on average more spatially mobile than workers in other occupations. The relatively privileged position of tech workers in an increasingly globalizing world has provided them with a rather positive outlook on international trade, migration, and diversity. Additionally, the international experiences of tech workers have also equipped them with schemas to deal with ambivalent social logics. Tech workers hold a specific form of sociocultural capital in their ability to rapidly adapt to new contexts—whether through contacts into different regional work contexts, language skills, or appreciation of regional culinary tastes. Tech workers are used to new situations whose elements are reshuffled time and again. This sociocultural capital grants them with a greater capacity to play by the logic of hybridity, and possibly even the logic of contradiction. Tech workers' active reconfiguration of their subjectivity must be seen in the context of this resource, which they were able to generate on the basis of previously existing forms of capital. An ease with hybridity, or at least certain forms of hybridity, then, is not an equally distributed character trait of modern subjects. Instead, it is something that must be discovered and learned in school, cultivated in the private sphere, trained in the academic field, and demonstrated in the world of work and beyond.

Hence, the capacity to authentically cultivate hybridity is also linked to class, gender, and race. Studies have demonstrated that the identity of the communicator is less open to individuals with working-class backgrounds as well as to People of Color (Amrute 2016, 67; Chua and Mazmanian 2020).

Furthermore, my research reveals that tech professions that have a relatively large share of women, such as UX design, are constructed one-dimensionally as predominantly communicative or creative but not technical (I will discuss the implications of this in depth in the following section). Thus, while certain poststructuralist theories universalize subjectivity as hybrid and fluid (e.g., Deleuze 1994; Deleuze and Guattari 1987), the empirical world reveals that authentically performing hybridity depends heavily on one's socio-structural background and position. Hybridity divides tech workers along the lines of class, gender, and race, and, in undermining their potential for collectivization, thus contributes to the contradictory logic of their class formation.

Finally, even tech workers' positioning vis-à-vis climate change cannot be understood without consideration of their biographies and resources. The climate crisis is an objective crisis, but it is also something of which subjects need to make sense. Sociologically, we must openly ask why and how tech workers, and the professional middle classes in general, signal more care about climate change than other classes (Huber 2022). One explanation lies in their predominantly middle-class backgrounds, which provide them with a particular set of experiences around managing limited resources. While members of the middle class do not have the abundance of resources enjoyed by the upper classes, they are also insulated from the scarcity of resources experienced by the lower classes, where life is more closely bound to necessity (Bourdieu 1984; Neckel 2017). The middle-class experience of managing limited resources corresponds to the way that the climate crisis is framed by scientists, politicians, and the media as a situation where we must use our limited natural resources more wisely. A further homology between tech workers' class position and the expression of their ecological habitus is found in their embeddedness in the knowledge economy and their skills in dealing with large datasets. My interviewees are academically trained in data analysis and distrust human decision-making when it comes to complex tasks. These two aspects are linked both to a belief in anthropogenic climate change and in the power of tech as a solution to the problem. Digital professionals hold a data-based form of cultural capital that increases the likelihood of belief in scientific reports about climate change, which generally are data-based (e.g., IPCC 2022). Furthermore, tech workers' data-based cultural capital also propels a paradox attachment to tech companies. Despite their critique of tech companies, tech workers ultimately still regard them as entities who through their data treasures might be capable of innovating pathways out of the big ecological crisis of our times.

5.4 PROFESSIONAL AND NATIONAL DIFFERENCES

The analysis so far has focused on accounting for the contradictory class formation of tech workers in the broader context of multiple crises as well as against the backdrop of their socio-structural positions and backgrounds. Now, the focus shifts to exploring the role played by *professional differences* and *national differences*. While this book has emphasized striking commonalities among data scientists and UX designers as well as between tech workers in the US and Germany, it has also revealed notable differences throughout, and these differences also have a significant part to play in shaping the internal dynamics of tech workers' class formation and the often contradictory ways they navigate their social worlds.

Professional Differences

One key element that undermines the formation of tech workers as a fully realized class has to do with the significant differences in status between professions in tech. As we have seen, the seemingly more "creative" and "qualitative" profession of UX design comes with lower average salaries and finds its roles symbolically devalorized by professions that self-classify as more "technical" and "quantitative," such as data science (we may expect that similar self-classifications as "more technical" are deployed by software engineers). Within the social worlds navigated by tech professionals, being seen as "more technical" is tied to better access to various resources. Constructing the task of designing digital apps as "less technical" than the coding of digital algorithms not only sustains a marginalization of women in tech (since women are more often located and pushed into seemingly "less technical" roles) but also generates divisions within the occupational segment of tech workers. By constructing some professionals as more genuine "tech" workers than others, the collectivization of workers in tech is obstructed from within the occupational segment.

From the perspective of the sociology of professions, this finding should not come as a surprise: Professions are traditionally theorized as privileged occupations that must constantly defend and valorize their task jurisdiction in a competitive "system of professions" (Abbott 1988). With regards to tech workers, however, it is important to highlight that these emergent digital professionals are typically embedded in corporations. Classical accounts of professions often envision these as somewhat entrepreneurial actors, such as in the case of independent doctors or lawyers (Parsons 1939). Tech workers, however, must be considered a case of professionalism that supports the diagnosis by Muzio et al. (2011) of a development in the world of work toward "corporate professionalization." Tech workers are typically employed by organizations and are not particularly self-reliant economic actors.

This position fuels contradiction because class interests and professional interests are difficult to align. In one sense, tech workers share a range of joint class interests they could pursue collectively with colleagues who also hold a highly sought professional expertise, vis-à-vis the managerial and entrepreneurial class within the organizations in which they are embedded. Together, data scientists and UX designers could pursue their shared interests and attempt to alter the broader power structures. Yet, in another sense, their embeddedness within a single corporate organization introduces a dynamic where it may seem more straightforward to secure more resources through horizontal struggles with other professions. In this scenario, data scientists and UX designers would compete with each other over jurisdictions and status within the broader power structures, challenging not the elite but rather their team members in different professional roles. Thus, corporate professionalization places tech workers in conflicting arenas of struggle, presenting them with two distinct pathways for advancing their positions.

Another key professional difference between data scientists and UX designers that contributes to their contradictory class formation is found in their differing levels of cultural capital. Data scientists typically have more institutionalized cultural capital than UX designers. Most of my interviewees had attained master's degrees, and six held PhDs. Among the twenty-six UX designers I interviewed, less than half held master's degrees and not one had a PhD. UX designers, on the other hand, appear to possess greater amounts of embodied cultural capital. In the interviews, they could typically talk about their worldviews, professional roles, and lifestyles in more elaborate and playful ways. While data scientists would be misunderstood as pure nerds, it remains that UX designers were a notch more talented in cultivating the identity of the communicator and along these lines also signaled more emphatic capacities. This professional difference can be interpreted against the backdrop of the relatively lower organizational status of UX designers. Excelling in communicative and emphatic skills might not only be necessary for their work (e.g., interactions with users) but also within the organization in the form of justificatory work that conveys the importance of their designs for the envisioned functioning of digital technologies.

With these findings in mind, it also becomes easier to understand why UX designers often presented a somewhat more critical mindset than data scientists about the tech industry and society in general. While the return of social critique is a social code that characterizes and unites tech workers, UX designers more frequently cultivate radical critiques, while data scientists remain somewhat more oriented toward voicing the need for incremental change. This difference can be said to contribute to their contradictory class formation in the sense that it makes a shared political vision more difficult to generate. While tech workers who lean toward the radical ideal type favor more encompassing societal transformations, such as those promoted by Bernie Sanders and supporters of "democratic socialism," others are content with what moderate democratic candidates have to offer and are rather at unease with more radical approaches. Surely, it is worth emphasizing that the overwhelming leftish political orientations among tech workers in both the US and Germany (where preferences for the "Green Party" dominate) create a space for political mobilization. On the other hand, it is important to keep the fine political differences in mind and how they correspond with more or less privileged professional positions within the tech industry.

There are, to be sure, further professional differences between UX designers and data scientists. In terms of leisure activities and tastes, data scientists tend to embrace ordinary lifestyles more strongly: They engage less frequently in high-brow cultural activities than UX designers. Mindfulness, though, plays a more significant role in the conducts of life of UX designers. However, I find that the central differences driving tech workers' contradictory class formation are rooted in variations in professional status, differing compositions of cultural capital, and divergent "leftish" political orientations.

National Differences

This investigation has consistently pointed to striking similarities between tech workers in the US and Germany. Their hearts and minds are very much in sync. Unlike other comparative studies of middle-class individuals across Europe and the US (e.g., Christin 2020; Lamont 1992), my study therefore does not mainly present a story of genuinely distinct national figurations. Two factors can be identified as contributing to the strong commonalities across the Atlantic. First, the US tech industry serves as a cultural role model for the tech industry in the Global North and beyond. Both tech corporations and tech workers in Germany are oriented in many, though not all, ways to the organizational cultures of the frontrunner of the digital economy. Secondly, the occupational segment of tech workers is comprised of a highly international workforce. In the US, twelve of my twenty-six interviewees reported being a foreign national; in Germany the figure was sixteen of twenty-six. This data roughly corresponds to more representative data collected by surveys (Baron 2018). Such an international makeup of the workforce means that national cultures are less pronounced, as compared to other industries where workers have grown up in the same nation state, and indeed for which career progression follows more distinctly national routes.

Nevertheless, notable differences exist between the two national sites. While tech workers across the Atlantic share a reorientation away from ideals of autonomy and creativity toward more other-oriented moral codes, there are sometimes differences in the content of this reorientation that will, in turn, constitute an obstacle for a possible transnational class formation of tech workers. One key difference is that tech workers in the US are more heavily concerned with avoiding (re)producing *bias* through their digital labor, while German tech workers are more heavily concerned with issues of *data privacy*. While the prevention of bias reproduction and data privacy violations may align in a lot of cases, at times, avoiding the discrimination of marginalized social groups through digital applications may necessitate collecting even more and especially more fine-grained data about individuals. Furthermore, the two moral issues may call for responses from different institutions. While data bias may manifest differently across cases and thus require more company- or even project-specific responses, the tackling of data privacy violations appears more suitable for broad-stroked governmental regulations (as manifested in the GDPR European data protection law).

The differences in moral codifications across Germany and the US generate further insights when placed in dialogue with findings by Di (2023). In her comparative study of tech workers' perception of big data ethics in the US and China, Di finds that tech workers in the US are more concerned with data bias while Chinese tech workers are more concerned with data privacy and worry about digital surveillance. At the same time, Di shows that tech workers in China are more oriented toward technosolutionism, which manifests in the sense that they are highly optimistic that big data applications can help to reduce economic inequality across different geographical regions. Thus, while German tech workers are aligned with US tech workers through their skepticism toward large-scale technosolutionist promises, they share (for different reasons) a pronounced concern for data privacy with Chinese tech workers. This comparison reveals that there exist *national varieties of social critique* among tech workers that must be taken into account in any reckoning with their potential for transnational organizing.

A second key theme that differentiates between tech workers in the US and Germany and contributes to their contradictory class formation has to do with *meritocratic legitimation*. To recapitulate, tech workers in both the US and Germany see their privileged (upper-) middle-class positions as heavily dependent on their individual meritocratic achievements. While they would be misunderstood as holding a naïve belief in equal life chances for everybody, they do cultivate a normative horizon in which life outcomes are firmly connected to individual merit. This belief already contributes to a contradiction in class formation since it contrasts with many of tech workers' more abstract social critiques. Tech workers, especially those with structural social privileges, rarely connect the categories of class, gender, and race to their individual lives and career outcomes. But there is another layer to this issue if we turn to the comparative dimension. While tech workers in Germany present meritocratic legitimations more often through recourse to "hard work," US tech workers add the notion of "talent" to the equation. In other words, when tech workers in Germany justify and make sense of their privileged life outcomes, they primarily highlight the hard work that they have put into their careers. In the US, hard work also plays an important role but is often accompanied by an emphasis on the role of talent—a justificatory scheme that was rarely mobilized by German tech workers.

As Friedman et al. note and discuss based on other research, claims to meritocratic achievements through hard work do not essentialize inequality, while claims for the relevance of talent imply "a sense of merit-scarcity that can be used to justify an almost infinite degree of inequality" (Friedman et al. 2024, 2). Thus, the different constructions of merit across the two settings suggest that tech workers in the US will be harder to win over for more structural reconfigurations of economic life, and more approachable for discourses that legitimize inequality based on a discourse around talent. In Germany, on the other hand, promoting solidarity without recourse to achievements or effort does not appear advisable. We can interpret these different meritocratic legitimizations across Germany and the US against the types of welfare states that tech workers are embedded in. While the US can be regarded as the ideal typical liberal welfare state, Germany is typically classified as a conservative welfare state with social services that valorize work over success (Esping-Andersen 1990; see also Rothstein 2022 on the comparison of Germany and the US).

Further national differences exist, which have been discussed in previous chapters. With regard to lifestyles, we have seen that US tech workers foreground an ordinary way of life and a distance from high-brow cultural activities to an even greater degree than German tech workers. Furthermore, it was shown that US academic institutions offering data science and UX design study programs place more attention on teaching communication skills as well as ethical thinking next to technical skills. Along these lines, it is also notable that US academic institutions offer more cooperations with tech firms (so-called industry partners). In terms of the contradictory class formation of tech workers, though, I hold that the outlined *national varieties of social critique* as well as the *national varieties of meritocratic legitimization* are the most powerful drivers of contradiction in the transatlantic class formation of tech workers. Zooming out, it has thus become clear that, in terms of their moral codes, US and German tech workers are tuned in to the same overall sound, if not always to the same track.

6 CONTOURS OF A NEW SPIRIT OF CAPITALISM

Extending Max Weber's seminal work on the protestant ethic, Luc Boltanski and Ève Chiapello's *The New Spirit of Capitalism* (2018), first published in 1999, articulates the history of capitalism as a story of ever-evolving "spirits" of capitalism. Weber famously argued that purely economic or technological explanations of the rise of capitalism put the cart before the horse (Weber 2005). It was the Protestant belief in hard work and an ascetic way of life, he argued, that made Western Europe, among all world regions, the cradle of capitalism. Boltanski and Chiapello radicalize this theoretical perspective. For them, the spirit of capitalism is not a one-time phenomenon: Unlike Weber, they do not consider cultural forces to lose relevance as capitalism matures. They suggest that capitalism not only emerged through cultural meaning systems but also continues to sustain itself through constantly evolving cultural belief systems that power and legitimize its operations. Building on this foundation, Boltanski and Chiapello propose a new explanation for the shift from the regulated Fordist epoch to the neoliberal, post-Fordist capitalist age. In a nutshell, the duo's *Zeitdiagnose* is that neoliberalism only became possible through the appropriation of the "artistic critique"—the demand for more creativity and autonomy within the world of work. By (symbolically) taking up this demand, capitalist actors found a powerful way of justifying the deregulation of the economy and the precaritization of employment conditions.

Boltanski and Chiapello's account aligns closely with analyses of the specific cultural logics underpinning digital capitalism. Most notably,

Barbrook and Cameron argued that the tech industry in the US Bay Area operates by means of a "Californian Ideology," a potent fusion of hyper-individualism and techno-futurism with neoliberal principles (Barbrook and Cameron 1995; see also Turner 2008). However, my empirical findings in chapters 2–4 challenge these established frameworks: Neither the notion of an artistic critique nor the concept of a Californian Ideology adequately captures tech workers' multifaceted cultural beliefs. While creativity, individualism, and market orientations remain significant forces, tech workers' social codes reveal a deep preoccupation with social issues, particularly economic inequality and insufficient diversity. My research warns that to characterize digital professionals as purely market-oriented individuals, whose ethics stop at expressive individualism, would fundamentally misread their reality. A profound, if contradictory, disillusionment with neoliberalism runs through the tech workforce. The animating spirit of the neoliberal, post-Fordist epoch appears to be waning.

This development, however, does not signal the end of capitalism's capacity to adapt to new cultural currents. This chapter reveals how the social codes of tech workers are appropriated and repurposed in the service of yet another spirit of capitalism, one that departs from the post-Fordist "artistic critique" identified by Boltanski and Chiapello to adopt a more explicitly *moral* orientation. Three new modes of appropriation and co-option emerge from my analysis, which is based on research conducted in the early 2020s. Each mode exploits the ambiguities within tech workers' codification principles in the given conjuncture. Together, they work to undermine tech workers' emancipatory potential, deepening the contradictions within their class formation. First off, tech companies attempt to infiltrate and re-reprogram tech workers' investment in social critique through a kind of *corporate humanism*. In the early 2020s, tech firms branded themselves as humanist actors, with their self-proclaimed pursuit of moral "initiatives" overshadowing their economic "projects." However, as we will see, buying into this branding came with hidden costs: It lended a fresh veneer of legitimacy to the tech elite while obscuring their underlying capitalist motives. Secondly, tech firms came to appropriate tech workers' hybrid subjectivity through a *play of differences*. This approach involved a superficial embrace of tech workers' valorization of diversity, which in reality sidestepped meaningful reform of workplaces or business models. The play of differences also extends to strategic translations of tech workers' professional ethics as well as the capitalist co-option of countercultural events like hackathons. Thirdly, tech firms also appropriated the lifestyles of their workers. Through the strategy of *compassionate design*, companies drew on tech workers' everyday habits of ordinariness and mindfulness to craft an image of the tech industry as an inclusive and nurturing environment where the needs of both employees and customers are valued and addressed.

Whether these schemes will endure remains to be seen as leading tech companies in the US appear to be shifting their ideological activities and corporate cultures in the context of closer partnerships with, and subordinations to, the US government and Donald Trump's second administration. Empirically, this chapter may therefore offer some rather historical accounts (at least with regard to the US tech industry). Theoretically, the chapter reveals capitalism's ongoing capacity to reconfigure market-limiting social codes into market-expanding ones by reconstructing how corporate cultures evolved and adapted to various critiques of neoliberalism—adaptations that have served to justify business operations and simultaneously exert cultural forms of control over the labor process.

6.1 CORPORATE HUMANISM

As we have observed, tech workers are actively developing a range of moral codes as they strive to transcend the bounds of neoliberal subjectivation. The new normative orientations, however, are in the process of losing their critical edge. A *corporate humanism* has taken root, enabling a cultural rewiring of tech workers' socially critical visions. This corporate humanism goes beyond "technosolutionism" (Morozov 2013) by advancing and fusing corporate repertoires of guardianship, social justice, and sustainability. The rhetoric of "projects" that was typical under post-Fordism (Boltanski and Chiapello 2018) has given way to a rhetoric of "initiatives," which reframes the concept of temporary in-company work groups within a narrative of values and engagement. Nearly all major tech companies, along with many start-ups, seek to portray their activities as motivated by significant noneconomic objectives. The initiatives they promote, though rooted in traditional Corporate Social Responsibility (CSR) frameworks are now increasingly intertwined with "Big Data" and technological tools, with the declared aim to maximize impact. Facebook, for instance, launched a "Data for Good" initiative, which positions the company as an active agent in responding to a range of societal crises (Facebook 2022). For example, in response to the Covid-19 pandemic, the company asserted that it had provided data to researchers and other stakeholders to assist with limiting the virus's spread. Facebook brands its unique data resources as essential tools for saving and protecting lives, though this promise comes with hidden costs. By framing its data power as crucial in combating crises, Facebook is able to deflect criticism of its status as a quasi-monopoly. The underlying message is clear: The more data a company holds, the more effective it can be in pursuing moral initiatives. This framing not only serves to legitimate Facebook's existing activities; it also opens up new markets, especially in the governmental sector. The strategy of corporate humanism not only pacifies the social critique of tech workers but actually transforms the market-restricting social code into a market-enabling one. To counteract this appropriation, tech workers—alongside activists, scholars, and other relevant actors—would need to critically examine whether the privatized data held by quasi-monopolous tech companies could be used more effectively for social purposes if it were a publicly accessible social good.

At times, tech companies also adopt a seemingly critical discourse on "artificial intelligence" (AI) as a way of engaging in heightened moral self-presentation and tapping into the ethical concerns of tech workers. Industry leaders, including Elon Musk, Bill Gates, and Sam Altman, have issued public warnings about the possibility for AI to develop a life of its own and pose an existential threat to humanity. In one case, Gates, Altman, and several academics and senior tech professionals cosigned a letter published by the Center for AI Safety, which declared, "Mitigating the risk of extinction from AI should be a global priority alongside other societal-scale risks such as pandemics and nuclear war" (AI Safety 2023). Such a display of corporate humanism on the part of contemporary tech firms builds on and but crucially surpasses, in orientation and scope, the established Corporate Social Responsibility doctrines historically embraced by capitalist firms (Hanlon and Fleming 2009). What emerges is not only a rhetoric of responsibility but a deeper logic of *guardianship*, through which tech companies and their leaders seek to portray themselves as custodians of humanity's future. Whether the personal belief systems of tech leaders or the future trajectories of AI actually align with these public statements is an open empirical question. However, it can be argued that such statements allow tech leaders and their companies to cultivate an image of themselves as reflective actors, guardians who watch over society in times of great technological, economic, and political upheaval. Moreover, by spearheading calls for AI regulation, tech leaders arguably strengthen their influence in any future regulatory process, increasing their control over emerging markets and protecting their dominant positions from up-and-coming start-ups in the AI landscape. Simultaneously, the discursive focus on doomsday scenarios diverts attention from the very real harms already being caused by AI in the present moment. These include the significant environmental damage associated with AI "innovations" as well as the troubling reality that even the most advanced digital technologies are complicit in reproducing, and at times exacerbating, systemic discrimination against marginalized social groups (Benjamin 2019; Chun 2021; Lehuedé 2025). Finally, the logic of guardianship arguably helped prepare the cultural terrain for the latest intensification in collaborations between tech leaders and the second Trump administration. This is because the logic of guardianship resonates with a more authoritarian approach to politics, as both legitimize a high concentration of decision-making power in the hands of select actors.

A final arm of corporate humanism in the tech industry consists in the growing number of charities and foundations that are created or financed by tech entrepreneurs (Creech and Parks 2022; Williamson 2018). An array of influential actors in the tech industry (and beyond) are channeling more or less significant proportions of their fortunes into organizations with an explicit and publicity-seeking social mission—which often comes with the benefit of tax reduction (Acs and Phillips 2002, 199). Tech leaders appear to feel the need to show some form of response to societal problems. This development may be interpreted against pre–World War 2 levels of economic inequality in the Global North, especially with regard to the unequal distribution of wealth (Piketty 2014; Waitkus and Groh-Samberg 2017). Through philanthropic actions, the moral transformation of the entrepreneurial self by tech workers seemingly finds a direct response from the entrepreneurial class, who feel compelled to present themselves as "giving back." Establishing charities and foundations can be interpreted as co-option specifically of the principle of solidarity. While the decision as to where tech billionaires will invest their substantial profits is clearly undemocratic, the move is usually justified, more or less explicitly, with the argument that philanthropists must step in where other actors, including governments, fail to deliver.

At this point, it is important to stress that the formation of a moral rather than artistic spirit of capitalism should not lead to a general devalorization of other-oriented motives and actions. Just as creativity, flexibility, or autonomy are values that should not be solely judged on the basis of their complicity with capitalism in one specific conjuncture, so, too, do solidarity, social justice, and societal responsibility enfold different forces in different historically specific orders. Furthermore, it seems ill-advised to consider all morally justified and morally motivated actions by tech companies (and their founders) as essentially condemnable strategies of domination (let alone ones led by omniscient or omnipotent strategists). This has become particularly evident in light of recent political shifts, in which forms of "green capitalism" or "progressive neoliberalism," which are now under pressure, may be superseded by an economic conjuncture that reverses even the limited progress made on some fronts in recent decades. While it is important to recognize ongoing and looming appropriations of social critiques, the situation is more complex than was the case with the artistic critique, as it is not self-oriented but rather other-oriented, prosocial principles that are being appropriated. A way forward may lie in placing particular emphasis on demanding a more meaningful and impactful enactment of humanitarian values by tech firms, and on advancing aspects of critique that remain unaddressed by capitalist enterprises, rather than dismissing their social agendas outright. With regard to tech workers, this concerns, in particular, their calls for a less venture capital–driven tech industry, their openness for more regulation of the economic field and distribution of wealth, as well as their sympathies for unions.

6.2 A PLAY OF DIFFERENCES

A second scheme through which tech companies advanced a new spirit of capitalism in the early 2020s can be referred to as *a play of differences*. While closely tied to corporate humanism, this scheme can be differentiated by its specific strategic effects. Central to this approach is the symbolic celebration of diversity. Despite the continued overrepresentation of white and Asian men in the tech workforce and ongoing reports of bias in technology (Chun 2021; Neely et al. 2023), tech companies invested considerable effort in appearing to value diversity and to show sensitivity around issues of discrimination in hiring practices and product development (at least before Trump's reelection in 2024 and his administration's offensive against DEI programs). The efforts included public commitments made through marketing campaigns, diversity reports, job ads, and official statements. Tech firms were eager, and still are to some degree, to present themselves as organizational habitats where people with all kinds of identities and backgrounds harmoniously come together to improve the world.

The heightened symbolic commitments to diversity allowed tech firms to cast themselves as progressive and inclusive employers, even when the reality of their workforce composition and organizational power structures remained largely unchanged. In other words, tech firms *capitalized on the moral currency of diversity and inclusivity*. This is often also the case with regard to so-called diversity mentoring programs or inclusion trainings installed by tech companies. Ultimately, though, diversity agendas seemed to have functioned primarily as a smokescreen, drawing attention away from the ongoing underrepresentation of marginalized groups and the continued inscription of bias in seemingly "cutting-edge" technologies.

To counter the co-option of diversity-focused critiques, it would be necessary to see tech workers (and other social groups) actively confront processes of appropriation. This would mean addressing instances where diversity is celebrated only in superficial or tokenistic ways. It would also mean challenging tech companies' tendencies to silo diversity efforts into select projects, embracing inclusion only as a tool for more marketable "innovations" that expand consumer reach. Equally important is scrutinizing the limited impact of many current diversity programs while advocating for the development of more meaningful and impactful alternatives.

Another important point to consider is that tech corporations largely turn a blind eye to issues of economic inequality. While we have seen firms place an immense (symbolic) emphasis on the dimensions of gender and racial inequalities, the economic dimension is often notably ignored. Though these dimensions of inequality overlap, they are far from the same. Provocatively put, the scheme of a play of difference in tech may reinforce a system of "progressive neoliberalism" (Brenner and Fraser 2017) where incremental progress in terms of gender and racial inequality serves as a distraction from heightened forms of economic inequality. In this way, the corporate co-option of social critique can, ironically, fuel a contemporary capitalist order that exacerbates economic exploitation. Moreover, when advances in gender and racial equality occur without parallel gains in economic equality, they may fuel backlashes against diversity efforts. To challenge this dynamic, tech workers would need to pair a less conciliatory concern for diversity with a more pronounced and forceful critique of economic inequality. Given that tech workers' concerns around diversity stem from a shared economic position, they are perhaps better positioned than most social movements to connect critiques of economic inequality with critiques of racial and gender discrimination and thus avoid unproductive internal disputes that often undermine advances.

Beyond the strategic co-option of diversity-related issues, a play of differences can also be identified in the tech industry's *endogenization of hacker culture*. Most clearly, this process is taking place through the countless "hackathons" sponsored or even hosted by tech companies. As various social scientists have pointed out, the culture of hacking can be understood as a critical legacy of virtual communities and computer enthusiasts (Dunbar-Hester 2019; Rone 2021; Turner 2008). In most hackathons, though, the critical or countercultural element of hacking fades into the background. As Irani (2015b) has shown, hackathons primarily function as capitalist spectacles in which teams typically compete with each other over developing a monetizable application in a relatively short time period (usually 24–48 hours). Hackathons celebrate a fast-paced and seemingly gamified capitalist production model—one that is unlikely to foster deeper ethical discussions. They hold the power to translate work-heavy "all-nighters" into a seemingly fun and exciting activity. Hackathons serve as a new kind of "organizational ritual" (Kunda 1992, 92, 178) to engineer culture in the interests of capital. While semantically feeding off the edgy and countercultural legacy of hacking, most hackathons nowadays are geared toward commodification and producing entrepreneurial citizens, thus leaving little space for a reflective engagement with differences.

Finally, we can also observe a play of differences in the way tech companies *rebrand the professional roles of their tech workers*. Engineers and designers in tech often perceive of themselves as caught between the distinct and conflicting interests of users and businesses, but tech companies actively work to reshape this frame. This involves spinning a narrative that reframes contradictions as challenges to be playfully engaged with. They cast their white-collar workers as professionals who can easily harmonize different needs, presenting this brokering as an opportunity for playful problem-solving. Through this strategic narrative, tech firms downplay the contradictions in workers' professional tasks, suggesting that user needs and business goals are not at odds. This generates the illusion of win-win solutions, masking the ethical complexities and power imbalances that professional tech workers are experiencing in their day-to-day work.

The symbolic appropriation of tech workers' concern for diversity, the endogenization of hackathons, and the translation of ethical professional struggles into playful, harmonious challenges all play a role in shaping yet another spirit of capitalism that is relatively distinct from the post-Fordist version described by Boltanski and Chiapello. Though autonomy, flexibility, and creativity remain present here, they no longer take center stage. The ideological forces of capitalism no longer appear preoccupied with addressing critiques that portray its systems as alienating. Instead, the leading industry of contemporary capitalism seems, or at least seemed, more eager to demonstrate that it can successfully tackle issues of social justice while creating work environments where dealing with complex moral issues is seen as a playful endeavor.

What makes this spirit of capitalism theoretically notable is how it emerged in a markedly distinct environment and at a different pace to previous iterations. The appropriation and co-option of the artistic critique, as described by Boltanski and Chiapello (2018), gained momentum only in the 1970s and 80s, well after the 1960s' widespread political protests and student revolts, which had posed a serious challenge to the capitalist system, particularly as they were bolstered by political parties advocating for market-restricting socialist agendas. The more moral and less artistic spirit of capitalism outlined in this book also emerged in a highly politicized environment, but one where socialist ideals have largely faded or been reduced to reformist visions that pose little threat to the dominance of market principles in the organization of the economy. Moreover, the outlined spirit of capitalism seems far more proactive, moving at a faster pace than its predecessor. Critiques of insufficient diversity were extremely swiftly integrated into capitalism through institutional channels. The rapid adaptation suggests that analyses of capitalist ideological activities must go beyond consideration of content to also take account of form and context. Like a snake that never sheds its skin at the same pace or in the same place twice, the spirit of capitalism continually adjusts its speed and movements to meet the moment. Hence, it may very well be possible that the reelection of Donald Trump as president in 2024 marks the onset of yet another conjuncture, where the still emergent moral spirit of capitalism is superseded by a cultural logic more attuned to a political-economic system built around de-globalization and increasing illiberalism. Only time will tell.

6.3 COMPASSIONATE DESIGN

But it is not only the social critique and professional ethics of tech workers that provide the nectar for a cultural and moral rejuvenation of the capitalist economic system. The lifestyles of ordinariness and mindfulness of the digital middle class must also be critically examined. Following the logic of capitalism's tendency to expansion, the meaning systems that these lifestyles rest upon may be rewired to propel commodification processes. In the following, I argue that both ordinariness and mindfulness may be appropriated through the scheme of *compassionate design*.

As we have seen, the lifestyle of ordinariness, which seemingly connects tech workers with social classes that hold less cultural capital, may also (unintentionally) serve as an ideological resource to legitimize inequality. An "ordinary" way of life can be interpreted as a strategy (without strategists) deployed by affluent actors to establish cultural legitimacy. Friedman and Reeves (2020) diagnose this mode among Britain's upper class. My research shows that this cultural lifestyle also exists among the new middle-class fraction of tech workers. I now want to stress that ordinariness is being mobilized both at the leadership and organizational level of tech corporations. Among the entrepreneurial class, an appropriation of ordinariness is instantiated in "no-collar dress codes" (Ross 2003), which tech leaders such as Sam Altman have perfected. In tech, it is more difficult stylistically to distinguish the owners of the digital means of production from those who sell their labor power to them. The relevance of this phenomenon should not be underestimated since traditionally class conflicts happened between stylistically quite distinguishable groups with starkly varying tastes in clothing (Bourdieu 1984). In the world of tech and increasingly in other fields (Ross 2003), very similar clothes are worn across the professional, managerial, and entrepreneurial class. This situation can be said to make collective organization more difficult. In the context of increasing inequality, the loss in visual differences between workers and capitalists is a loss of class power on the side of labor.

Moving to the organizational level, we can also locate a strategic use of ordinariness in the architecture of office buildings, which, unlike the skyscrapers of the era of industrial capitalism (and indeed of the organizational self), are not intended to appear spectacular. Instead, the offices of start-ups are usually located in authenticity-signaling brick buildings, while the corporate headquarters of tech companies are usually only a few stories high and aim to signal flattened hierarchies and organizational transparency. Not just the dress codes but also the entire habitats of the tech industry are designed in a fashion that aims for easiness and openness. This includes the design and furnishing of office spaces, which are not configurated in the intimidating and slick way that is typical of the finance industry but rather intended to signal communality (see already Kunda 1992, 192; Müller 2023). Instead of cubicles or closed rooms, these offices feature open spaces, flexible workstations, and glass-walled conference rooms. Through this compassionate design of work habitats, which is calibrated toward signaling inclusivity and an absence of rigid hierarchies, it arguably becomes harder to perceive and especially oppose the operations taking place inside as exploitative or harmful for societies and individuals.

Furthermore, we can locate an element of ordinariness in many tech companies' claim to produce digital technologies that anybody can easily use. The self-branding of tech companies as product suppliers not just for technically skilled groups but also for ordinary people allows tech companies to appear sympathetic and warm-hearted while enlarging their base of potential customers. This strategy has been perfected by Apple, who brands its products as easier and more intuitive to use than those of its rivals, in particular Microsoft (Burgess 2012). There is a value in the value of ordinariness. This is especially the case for companies whose business models build more on purchases by individual customers than those of other businesses. Ordinariness can thus be located across different levels of corporations, including its business models, corporate culture, architecture, and dress code. Across these levels, ordinariness is rechanneled for purposes of cultural legitimacy and commodification that undermine the critical edge of its cultivation in the self-understandings of tech workers.

The lifestyle of mindfulness also carries the potential for appropriation and endogenization through a seemingly more compassionate form of capitalism. At first glance, by prompting a redifferentiation between work and life, the ethic of mindfulness appears to be a practice of decommodification and a technique to counter "social acceleration" (Rosa 2013). However, this new ethic can also function as a commodifying technique in the way it is leveraged to heighten performance at work. Many tech companies cater to tech workers' mindful technologies of the self by offering yoga classes or massage therapies at the workplace. Additionally, tech headquarters are often designed for maximum convenience, equipped with on-site childcare, pet care and laundry facilities. Buses operate from early morning until late at night, and cafeterias serve three free meals a day. The offering of these seemingly generous perks can be understood as a strategy to keep their employees in the workplace for longer and to improve their capacity for concentration. The seemingly humane worlds of work that tech companies advocate show contours of "total institutions" (Goffman 1961). Furthermore, the discourse around the technique of mindfulness risks individualizing mental illnesses or occupational phenomena such as burnout, as well as their proposed solutions (Purser 2019). If the practice of mindfulness is not complemented by structural changes in the organization and valuation of work, burnout can be dismissed by employers as the result of insufficient self-care rather than unsustainable work pressures.

A second way in which mindfulness may be appropriated and reconfigured to serve the economic interests of tech firms is through its strategic use as a quasi-religion. In her account of the tech industry, Chen (2022) has stressed how tech companies are striving to constitute habitats for community, purpose, and transcendence. In so doing, they are not only trying to make their workers stay around longer but also to bind them to their operations as believers and dedicated followers. To do so, they not only offer perks and a communal work culture but also a range of spiritual events, including yoga lessons and ostensibly "Buddhist" mindfulness sessions at the company's expense. Especially in the US, where religion plays a more important role in society than in Germany (Pollack and Rosta 2017), mindfulness can serve as a gateway for tech companies to tap into the religious needs of their workers and establish unique affective binds with them. Such binds may further hinder their orientation away from capital toward labor.

Together, corporate humanism, a play of differences, and compassionate design offer insights into the formation of yet another spirit of capitalism. The three cultural forms suggest that the sunset of the Californian Ideology does not signal the end of capitalism's cultural adaptability. Instead, they reconfigure tech workers' social codes to justify contemporary capitalist operations and enable further commodification.

The alluring cultural force of these schemes is a vital factor in understanding why highly educated, critically minded workers continue selling their labor power to hyper-capitalist tech firms. While this book does not want to dismiss the market-limiting potential embedded in tech workers' social codes, it is important to recognize that, when appropriated, these same codes can also serve to enable market operations. Further research is necessary to explore these dynamics and to trace the broader ideological shifts of contemporary capitalism. It is also crucial to stress that capitalist firms do not rely solely on ideological tools to advance their interests. They also employ noncultural forms of control, such as digital surveillance or open retaliation against union organizers (Kapoor et al. 2022). Still, the contours of a virtue-oriented, playful, and compassionate spirit of capitalism highlight how power and control are exercised in understated ways, with the potential to rewire self-understandings, further supporting this study's central claim that tech workers are engaged in a contradictory class formation.

7 GLITCHES IN THE CLASS MATRIX

In chapter 5, we saw how the social codes of tech workers signal their formation as a contradictory middle-class fraction. These workers cultivate a shared spirit of emancipation, yet remain entangled with certain neoliberal principles that obstruct class cohesion and undermine collective action. Zooming out, this formation was analytically situated in the context of multiple crises as well as tech workers' forms of capital and socioeconomic backgrounds. Chapter 6 built on this foundation, identifying signs that tech workers' social codes are increasingly drawn into the orbit of yet another spirit of capitalism by means of the schemes of corporate humanism, a play of differences, and compassionate design, which appropriate and reprogram tech workers' market-limiting social codes.

This penultimate chapter weaves together these insights from the previous two chapters to discuss the wider implications of the subjectivation of tech workers for the class system. In particular, I suggest that tech workers can "glitch" the class system through their *double contradictory class location*. By building on the empirical findings from interviews with tech workers, and drawing on Erik Olin Wright's theoretical concept of the contradictory class location (1976), we will see how tech workers occupy contradictory spaces both in economic terms, where they find themselves in between labor and capital, and in cultural and moral terms, where they are placed somewhere between emancipation and domination. The concept of the *glitch* proves useful to better grasp the nature and implications of this location. Traditionally, the glitch has been associated with gaming, where it is used to describe technical irregularities and flaws in games' source code that players can exploit (Bainbridge and Bainbridge 2007). In the context of this study, the glitch offers a metaphor for how tech workers, through their double contradictory location, can take advantage of strategic positions and openings within the class matrix, potentially reconfiguring the rules of the game. By inhabiting this double location, I argue, tech workers are able to widen the space of possibilities and engage in new alliances and "class blocs" (Gramsci 1971).

This chapter's second section explores how this double contradictory class location intersects with broader *class conflicts* in contemporary Western societies. In dialogue with theories about conflicts between the "new middle class" and the "old middle class" (Reckwitz 2020, 2021) or "cosmopolitans" and "communitarians" (Koopmans and Zürn 2019), my work positions tech workers at the forefront of a development where moral capital gains currency. The "glitchy" position of tech workers, we will see, has the potential to both bridge and exacerbate the divides and conflicts within Western societies' middle class depending on how the space of possibilities is maneuvered. Expanding the lens to a global scale will then allow us to consider the class formation of tech workers within the *global class matrix*. The spread of digital hubs and the employment segment of tech work beyond the postindustrial economies of the Global North (e.g., Irani 2019; Liu 2022; Shakthi 2023a; Takhteyev 2012) carries profound implications for global relations of class that warrant closer examination.

7.1 GLITCHING VIA A DOUBLE CONTRADICTORY CLASS LOCATION

My conceptualization of tech workers occupying a double contradictory class location draws on Erik Olin Wright's seminal work on class. Today, Wright is best known for having revitalized class analysis through quantitative methodologies (Wright 2005). However, another important contribution of Wright's was his differentiation of class locations (Wright 1976, 2004). Wright argued that there is an internal differentiation among those selling their labor power to make a living, structured by their relative positions within hierarchies of authority and the distribution of skill sets. Developing a new theoretical vocabulary, Wright argued that managers, supervisors, and professionals occupy a *contradictory class location*—managers and supervisors especially on account of their elevated position in authority hierarchies, and professionals on account of their codified expertise and control over their labor process (see also Abbott 1988; Freidson 1988; Pfadenhauer 2003). Building on this heuristic, I hold that tech workers generally hold an economically contradictory position due to their highly complex and pronounced professional expertise but also due to their notable command of authority. While they do not own significant shares of the means of production, tech workers have established a knowledge and habitual skill set that is both codified and (still) highly demanded on the labor market—granting them a significant degree of control over their labor, and providing them with substantial bargaining power. Furthermore, tech workers exert authority through their inscription power and the programming and designing of digital technologies that control and surveil other groups of workers, such as gig and crowd workers. Finally, tech workers sometimes take up elevated positions in authority hierarchies—even though my study focused mostly on mid-level professionals who were not usually in supervision roles.

Wright's concept allows us to get a better account of the economic class position of tech workers. However, there is more ambivalence at play, which we can grasp by systematically studying cultural, moral, and political positionings and avoid granting a primacy to economic relations. Extending Wright's (1976) theoretical notion, I hold that tech workers occupy a *double contradictory class location*. The location of tech workers is not only contradictory with respect to their positioning in the production process. Building on the previous chapters, I argue that the economic grouping of tech workers is also contradictory with respect to their ambivalent cultural and moral positionings that, as we have seen, are double-edged and prone to appropriation and co-option by their employers. In other words, the subjectivity of tech workers, which is structured by relatively autonomous cultural processes and is not determined by their contradictory economic location, opens a space for emancipatory politics as well as for the reproduction of relations of domination.

In making this conceptual contribution, my study also departs from Ben Tarnoff's theorization of tech workers by stressing ambivalence in subjective orientations. Tarnoff's pioneering work on tech workers suggests that those situated in a contradictory class location can "focus on the ways in which they are bourgeois, and identify with the capitalist class; or they can focus on the ways in which they are proletarian, and forge alliances with the working class. The tech worker movement offers a fascinating illustration of the latter phenomenon" (Tarnoff 2020). My study puts forward a different account of tech workers' identity, which stresses the structural ambivalence of their social codes. The epistemic doxa of contemporary capitalism is challenged by the economically intermediate grouping of tech workers in a fashion that makes alternative conjunctures thinkable, at the same time as they make possible a paradoxically dynamic stabilization of capitalism through the reproduction of certain neoliberal scripts. In this sense, there appears to be no antinomy between capital and its resistances (Amrute 2014, 108; Foucault 2007, 355).

I now wish to shed light on the emancipatory potential that arises from the double contradictory class location of tech workers using the concept of the *glitch*. In gaming communities, this term is used to describe irregularities in the source code of games that players can take advantage of—for instance, when deploying tricks to gain unusual abilities or to enter spatial areas of games that are not supposed to be entered (Bainbridge and Bainbridge 2007). Glitching can be understood as a movement that allows actors to explore interruptions of a system to create something new (Prior 2008). Transporting this concept to class analysis, I argue that the double contradictory class location of tech workers can be interpreted as allowing for the glichting of social relations. Similar to the capacities of figures such as the "marginal man" (better put: marginal self) in urban space who connects distinct or even antagonistic milieus and quarters (Park 1928, 881), we can understand glitching as the circumventing and undermining of symbolic and social boundaries. Specifically, with regard to tech workers, I hold that their double contradictory class location equips them with the capacity to access different class blocs that are separated from one another. In this way, tech workers carry the potential to establish connections across class fractions. Furthermore, as a central carrier group of an emerging spirit of capitalism, tech workers could use their positioning to negotiate the process from below and possibly even to mobilize other groups in resistance to it. Spirits of capitalism are dependent on spirits of emancipation, but they only superficially address critiques and instead focus on co-opting them to allow for heightened commodification. To prevent critiques and value systems from rejuvenating the capitalist system once again, both the exploitation of labor and of cultural repertoires requires challenging, and effectively pursuing this will require class coalitions.

Thinking social space in terms of a cyber space that includes glitches invites us to contemplate how unusual class positions may serve as vehicles for new class strategies. In the case of tech workers, specific possibilities for glitching are opened by the nature of their work tasks as well as by their command of authority through inscription processes. As discussed, the digital technologies designed and maintained by tech workers are instrumental for processes of control and surveillance. Various studies on gig and crowd workers have demonstrated how the labor process of these low-paid groups of digital workers is subject to new forms of algorithmic control (Altenried and Niebler 2022; Wood et al. 2019; Wood 2020). For instance, gig workers (who are part of the growing low-paid class of service workers that is increasingly superseding industrial workers) are confronted with apps that constantly monitor their delivery routes and may sanction workers for breaks or denied deliveries. Other groups of digital workers, including YouTube content producers (Niebler 2020), may be more constrained by algorithmic recommendation systems. Lately, high-paid professional groups outside of the digital economy, such as doctors or lawyers, have also been impacted by multifaceted algorithmic systems (Kellogg et al. 2019). Given these multiple levels of inscription power, tech workers would be a crucial grouping to be won over for class alliances. Apart from joining other groups of workers in social struggles, including strikes or other union-led actions, tech workers may seek to glitch the contemporary class matrix by developing technologies that minimize the exploitation of workers and open a greater space of possibilities.

According to Braverman, workers en masse have witnessed a decline in their command over the labor process in the twentieth century (Braverman

1974). The capitalist class wields power by stripping supervised workers of skilled activity. Managers and engineers, however, have been able to increase their command over the labor process, which Braverman considers a direct result of the dispossession of supervised workers from knowledge and craft. Tech workers mostly belong to the latter group, which has also been dubbed the "professional-managerial class" (Ehrenreich and Ehrenreich 1979). They develop the digital technologies that surveil and control workers in the digital economy and beyond. Paradoxically, tech workers may even develop digital technologies and services that control and deskill their own occupational segment (Steinhoff 2021, 214). This places them in yet another contradictory class location. While the so-called labor process debate has highlighted how the capitalist class and its allies can enfold power through the deployment of technologies that allow for control and deskilling, the case of tech workers highlights the fragility of this power, which relies on workers who develop those technologies and who may be impacted themselves by them. Given what we know about the social codes of tech workers, it is possible to imagine their opening a space of possibilities by not only refusing certain work tasks but also by building powerful alliances and coalitions between deskilled workers and upskilled workers. Just like a virtual matrix, every capitalist conjuncture offers spaces for glitching in the sense that taking advantage of specific locations in the game allows actors to generate unusual abilities.

Ultimately, workers are confronted with structural dynamics that aim to rewire their ambivalent social codes, but they also have the capacities to play and leverage their contradictory positionings. Their specific class location, work tasks, and subjectivation all hold potential for emancipation as well as for the reproduction of relations of domination. Both dimensions are present and entangled in tech workers' space of possibilities. Neither techno-optimistic accounts of tech workers, which dominated scholarly debates in the early phase of the internet (e.g., Hardt and Negri 2000), nor techno-pessimistic accounts, which have shaped scholarly debates since the consolidation of the commercial internet (e.g., Morozov 2013), are adequate to understanding and grasping their trajectory in social space. The pendle may swing out to both sides in different situations and depending on specific maneuverings as tech workers' codes invoke market limitations and market expansions, openings and closings of contingency, the building of bridges as well as boundaries. In the following sections, I will develop this theorization by discussing tech workers' glitchy class position in the context of debates on middle-class conflicts in the Global North and, subsequently, with respect to the global class matrix.

7.2 MIDDLE-CLASS CONFLICTS

In the twenty-first century, the analysis of the middle layers of societies in the Global North has shifted emphasis away from their relative stability and coherence (Schelsky 1967) toward processes of decline, fragmentation, and antagonization (Nachtwey 2016; Verchere 2017). A prominent example positing increasing antagonization is found in the work of the sociologist Andreas Reckwitz, who argues that there is a trend across Western postindustrial societies toward a conflictive constellation between a *new* and an *old middle class* (Reckwitz 2020, 2021). The new middle class can be characterized as the rising bloc within the middle layers, whose members generally hold academic degrees and tend to work in the knowledge economy or related industries. This bloc fundamentally distinguishes itself on account of its high cultural capital (Reckwitz 2021, 47). Despite a broad range of incomes, members of the new middle class are bound together via highly valorized cultural repertoires and practices that gained prominence in post-Fordism. They cultivate creativity, live in urban rather than rural areas, and embody cosmopolitanism. Furthermore, they are heavily engaged in practices of "doing singularity," seeking a life that has not been chosen "off the rack" (Reckwitz 2021, 49; see Baert 2022 for a critical discussion of Reckwitz' conceptualization of singularity). By combining economic success with culturally distinct self-fulfillment, the new middle class represents a fusion of bourgeois life and romanticism. The old middle class, on the other hand, can be considered the declining bloc within the middle-class segment. Members of this class work in occupations that are less academic such as traditional industry jobs, mid-level office and administration jobs, or independent craftsmanship. They tend to live in smaller cities, towns, or rural habitats and to consume standardized goods. The guiding codex of their lifestyle is not singularity but a concern for material wealth and a cultural framework that values order, ordinariness, and community (Reckwitz 2021, 52). Politically, the old middle class leans more, but far from comprehensively, toward conservative or even right-wing political parties. This big-picture account of the middle class, which leaves out a lot of nuances, corresponds in a number of ways to other panoramas that have gained prominence in the social sciences and beyond. Arlie Hochschildt, for instance, diagnoses conflict and estrangement in the US between globalists, on the one hand, and "strangers in their own land" on the other hand (Hochschild 2018). In political science, the theory of a confrontation of a new and old middle class is echoed in many ways in debates about a new cleavage between cosmopolitans and communitarians (Koopmans and Zürn 2019).[1]

How do tech workers fit into these broad-stroked accounts of middle-class dynamics? At first sight, tech workers would seem to clearly belong to the cosmopolitan and globally mobile new middle class. Tech workers show a strong preference for urban habitats and are experienced in international mobility. They consider new habitats and new people not as threats but as spaces for self-discovery. Furthermore, tech workers are a distinctly academic formation, with almost all my interviewees holding university degrees, often from elite institutions. They are indeed actors who combine economic success with exclusive cultural factors around education, internationalism, and intellectual curiosity. On closer consideration, though, it is possible to see how tech workers are also carriers of social codes that would allow them to "bridge" toward the more communitarian, old middle class. Especially through their multi-faceted moral codex, tech workers may glitch the cultural polarization in the middle layers of society. They value workplace solidarity and are sympathetic to the formation of unions and other collectives that may stretch across middle-class segments. There is an honest will among tech workers to replace the ideal of the self-oriented entrepreneurial self with the ideal of a more other-oriented moral self. So, even though tech workers' social critique is ambivalent and double-edged, it is important to highlight the potential for moral bridges.

We can also identify the code of ordinariness as working across social and symbolic boundaries within the middle class. In contrast to works that consider creativity or singularity as leading lifestyle principles of the academically trained middle class (Florida 2002; Reckwitz 2020), I find that knowledge workers in the internet economy place more importance on cultivating ordinary tastes and leisure activities. While my interviewees engage in exclusive cultural activities, their self-understandings and self-presentations are very much calibrated by the will to be down-to-earth. We may even speculate that the declining value of extraordinariness among tech workers stems from their familiarity with the influence of algorithms on everyday life, which likely disenchants notions of individual authenticity in one's way of life. In any case, tech workers do not actively seek to make their lives a work of art. Their ethics of existence challenges aesthetics of existence to a significant degree. And while the morals of inclusiveness and ordinariness hold the potential to shield tech workers from critique while they pull away economically, it is equally important to consider the emancipatory potential that is opened through this codex. In other words, just as with a glitch in a video game, the social critique and lifestyles of tech workers open a space of contingency within the class matrix.

Zooming out, it may tentatively be argued that the overarching movement behind these developments is a *shift from cultural to moral worth* in class formation.[2] At least within the digital middle class, cultural capital is losing exchange value on the trade floor of social currencies. This reshuffling of the determinants of social worth goes hand in hand with increased potential for new class alliances. Since the cultivation of moral schemas does not necessitate large amounts of cultural or economic capital (two resources that are highly secured by the upper echelons of society), its potential for boundary-makings (and -crossings) can, in principle, be more easily deployed by the lower classes. Class certainly shapes what kind of values we incorporate (Sayer 2005), but, especially as one grows older, these values may still be easier to adapt and modify than one's educational profile.

At the same time, the orientation toward morality within the middle class could also exacerbate intra-class conflicts within the middle layers of society. Questions around migration, in particular, may prove to be divisive insofar as opposing political stances on the issue lay claim to other-oriented moral principles (Beck and Westheuser 2022; Dörre et al. 2018; Mau et al.

2020). It's generally important to recognize that moral codification processes can be found across the political spectrum. Moral codes have also always been an instrument of right-wing actors, as they mobilize them to frame exclusionary politics as justified to protect "one's own". Finally, moral codes can exacerbate social divisions by becoming intertwined with cultural capital. For example, it can be socially required that moral claims be articulated through certain academic concepts or abstract terminologies. However, I consider it just as important to highlight the potential for new class alliances to be brought about by the shift toward moral worth in the dominant social codes of social critique and ordinariness. The growing relevance of moral worth establishes glitches in the contemporary class matrix that may be configured so as to provide new forms of organic solidarity across different occupational groups.

But how far does the growing relevance of moral worth within the middle class currently travel? Does the moral quest of tech workers, and the (occasional) unionization efforts among these, echo a larger development across major economic sectors? Studies of bankers, another elite fraction of the academic middle class, indicate a growing presence of ordinariness, although this has not been accompanied by social critique in their field (Neckel et al. 2018). In other sectors, though, we may see a corresponding shift toward social critique. In 2022, massive protests concerning pay and working conditions have taken place by teachers and nurses in the US and Germany (Sainato 2022; Ullrich 2022). This points to political action taking place across different segments of the middle class. In 2023, protests in the world of work have even intensified in the US. According to *The Economist*, 2023 saw the US private sector undergo the biggest strike wave since the 1980s (*The Economist*, 2023). In addition to 146,000 unionized workers in the automobile industry (most of which can be considered members of the old middle class) walking out in September 2023, 190,000 other workers, including Hollywood actors and screen writers (closer to the segment of the new middle class while at times leaning into the precariat), were on strike.[3] There appears to be a significant increase in collective action within the world of work—one that is driven by new moral orientations. Of course, new moral evaluation schemes will not manifest as a linear or homogenous development. Rather, it seems that morality is shaping the class matrix in different ways and to different degrees. Tech workers can be considered to take up a key position in this kaleidoscopic development given their combination of upper-middle-class volumes of economic capital with a moral and ordinary conduct of life as well as their crucial role in shaping the digital technologies that permeate social space. We can expect that the playing out of their contradictory class identity will ramify well beyond their professional field.

7.3 GLOBAL CLASS MATRIX

So far, we have seen how the social codes of tech workers translate into the formation of a middle-class fraction that carries the potential to reshape labor processes, support more precarious groups of workers, and bridge some of the middle-class divides in Western postindustrial societies. But how can we make sense of tech workers in the context of the global class matrix? To begin addressing this question, it is important to recall that most of my interviewees were not born in the countries where they currently work. The cadre of tech workers in the Global North is a highly international labor force. Globalization very much leaves its mark on national tech hubs. Despite the heterogeneous national backgrounds, though, I found that there is a common set of social codes that are recognized and shared. Indeed it can be argued that tech workers across Germany and the US show elements of transnational class formation (Sklair 2001).[4] If a fully formed transnational class would require a set of specifically institutional bonds, there nevertheless exists a certain set of schemas that are shared across the two sites. Tech workers' economic careers are international in scope, with a focus not on the nation-state but on spaces of innovation and habitats for a comfortable and moral middle-class way of life.

However, this transnationalism can assume very different forms. In line with Amrute (2019, 60), I found in my interviews that the lives of foreign tech workers, especially those from the Global South, are often heavily shaped by racialization processes. Due to the effects of temporary work permits, for instance, foreign tech workers must deal with greater levels of career uncertainty.[5] Furthermore, they are confronted with excluding subjectivations within the field that not only make it hard to "get in" but also to "get ahead." There is thus an institutional mechanism of "differential inclusion" (Mezzadra and Neilson 2013, 131), which generates different working conditions among workers and fractures potential solidarity among them (Amrute 2016). Furthermore, important regional differences exist. Despite pronounced similarities between the US and Germany, it is notable, for instance, that tech workers in the US are more concerned with data bias, while tech workers in Germany are more occupied with avoiding violations of data privacy in their work.

But making sense of tech workers in a global perspective also requires looking beyond the postindustrial economies of the Global North. In many parts of the Global South, vibrant tech industries are emerging. The Chinese city of Shenzhen, for instance, has become a giant tech hub that offers employment for both software and hardware production. As one of the special economic zones of China, Shenzhen plays a key role in the nation's ambition to develop a digital ecosystem that can compete with, and even technologically overtake, US big tech companies. Furthermore, India is home to a well-established industry for both high- and low-paid digital labor. From the 1990s onward, an IT boom has taken place in India, in which careers as coders and software programmers have become available especially to young, upper-caste men, while lower-caste individuals have often been pushed into lower-paid tasks, such as in the call centers (Gupta 2019, 123; Shakthi 2023a). Several countries in Africa and Latin America have also developed tech hubs. Nairobi, for instance, has been branded as "Silicon Savannah" due to its thriving start-up scene. In Latin America, lively tech ecosystems have emerged in cities such as Buenos Aires or Quito. Given this distribution, we may speak of "spikes" in the spatial structure of the global tech industry (Florida 2010).[6] There is now a centralization of the production of digital technologies within globally dispersed tech hubs (Takhteyev 2012).

Beyond the economic dimension, more research will be necessary to grasp whether the sharing of social codes among tech workers in the US and Germany extends to these hubs in other parts of the world. Existing studies indicate similarities as well as differences with my findings, even though such conclusions are highly preliminary given different theoretical frameworks and methodologies. One study that indicates a post-entrepreneurial subjectivity existing beyond the Global North is Di's examination of Chinese tech workers. Di's work highlights pronounced ethical concerns among these with regard to the application of big data in their work. Liu (2022) offers another study indicating social critique among Chinese tech workers, focusing on their activism in response to industry sexism. At the same time, Liu observes that Chinese tech workers face greater governmental suppression than their counterparts in the Global North as strikes and walkouts, for instance, are banned by the government and organizers face prosecution. Nonetheless, collective activism is possible as illustrated by Tan and Weigel's (2022) analysis of how Chinese tech workers have protested the overwork culture in their industry in the context of the so-called anti-996 or 996.ICU movement. This initiative saw Chinese tech employees push back against the expectation to work from 9 a.m. to 9 p.m., six days a week. Using the code-sharing platform GitHub as a tool, tech workers built solidarity and compiled and circulated rankings of firms based on their required work hours. Notably, American colleagues supported their efforts, urging Microsoft to resist any potential pressure from the Chinese government to shut down the GitHub activities. This example shows that transnational mobilization of tech workers across the Global North and South is possible, even if, as Tan and Weigel note, the protests ultimately did not succeed in bringing about significant reforms to the 996 work culture in Chinese tech companies (Tan and Weigel 2022, 221).

Other studies of tech workers in the Global South are also relevant for exploring potentials and obstacles for transnational class makings. Takhteyev (2012), for instance, analyzes tech workers in Brazil and conceptualizes them as "peripheral practitioners" who must align their products with technological expectations that are culturally shaped by dominating tech hubs within the Global North. Tech workers in the Global South are embedded in cultural and organizational contexts that are distinct to those of the Global North but must still play by some of their rules. Cultivating specific, even if differently accentuated, forms of hybridity thus appears to be necessary both for tech workers who have migrated to the Global North (as highlighted in chapter 3) and for those who work in the Global South. Such subjectivation processes are domination effects of a hierarchically structured global tech economy. Without romanticizing these subjugations, such instances may also be interpreted to provide some ground for cross-national bonds in the sense that homologous experiences are made within the occupational segment of tech workers across large spatial distances.

Another important study on tech workers in the Global South comes from Lilly Irani. Irani's research paints a vivid picture of the desire felt by tech workers in India to contribute to general societal well-being—a yearning that is rather similar to the sentiments expressed by my interviewees. However, Irani's work also illustrates in detail how this quest for the greater good in India is heavily intertwined with patriotic and nationalistic political frameworks (Irani 2019). This suggests that not only the state but also national cultures can potentially impede the making of transnational class solidarity. Moreover, Irani's study reveals that, in general, tech workers in India show pronounced and explicit openness and sympathy for entrepreneurial forms of subjectivity, a finding that is echoed in studies by Shakti (2023a, 2023b) on the Indian IT sector. This deviates from my empirical findings in the contexts of Germany and the US, indicating variations in tech worker subjectivity.

Finally, I wish to explore tech workers' position in the global class matrix in the light of regional differences in the trajectory of the middle class itself. It remains clear that tech workers constitute a rising middle class across the Global North and South. But the opposing trajectories of the middle class in these two regions, I hold, places tech workers in a further glitchy class position. In the case of the Global North, most scholars consider the middle class to be in a state of decline. According to a study by the Bertelsmann Group (Bertelsmann Stiftung 2021), the amount of people with 75 percent to 200 percent of the median income has decreased in Germany in the last two decades. In the US, this development has arguably been even more pronounced (Putnam 2016). Thus, despite the rise and growth of a few middle-class fractions such as tech workers or consultants (Schmidt-Wellenburg 2014), who have been considered members of the "new middle class" (Reckwitz 2021), the bulk of the middle class in the Global North is declining in terms of economic capital.

Unable to build up assets and so reliant on earnings, many middle-class subjects have come under pressure in the light of stagnating wages (Waitkus and Groh-Samberg 2017). Tech workers are one of the few fractions that is exempt from this general trend among the middle classes in the Global North. Recent layoffs appear to have been a correction to the hyper growth of the occupational segment during the Covid-19 pandemic. Tech workers remain one of a few middle-class segments in the Global North to have grown and prospered in the past 20 years. Expanding our scope toward the Global South, though, the situation of the middle class appears in a different light. In many Asian, Latin-American, and African countries, the middle class is rapidly expanding (Milanovic 2016).[7] Countries in these continents have undergone the transformation from agrarian into industrial societies rather quickly. Very roughly speaking, countries like China and South Korea but also Chile or Kenya have experienced radical transformations such as industrialization, urbanization, differentiation, or individualization within a time span not of two hundred but rather forty years. The global class matrix is thus structured by simultaneous yet distinct trajectories of the middle class.

This constellation places tech workers in a unique class position. At the global level, they constitute a growing class fraction across rising and declining middle classes, across flourishing and stagnating capitalist economies. In the Global North, tech workers are pulling away economically in an increasingly precarious society, while in the Global South, they are flowing with the bulk of society in an economically upwardly mobile direction. What results specifically from this positioning is hard to extrapolate. What is certain is that their glitchy position will allow tech workers as a global occupational segment to collect experiences of varieties of global capitalism from relatively secure economic as well as technologically influential positions, and thus with a good degree of leverage. Their key role within the world's leading economic industry grants them significant occupational power as well as the potential to play an important part in a potential global labor movement. Tech workers are packed in a speedy elevator within a multipolar global class matrix. They are a middle class in the making, and it remains to be seen on which floor they will get off.

8 CONCLUSION

> Not ideas, but material and ideal interests, directly govern men's conduct. Yet very frequently the "world images" that have been created by "ideas" have, like switchmen, determined the tracks along which action has been pushed by the dynamic of interest.
>
> —Weber (1946, 280)

In the late nineteenth and early twentieth century, telephone communication had to be manually facilitated by switchboard operators whose job consisted of connecting calls by physically plugging wires into the correct jacks on a central board (Ritzer 1971, 10). The occupation of the switchboard operator was automated from the 1920s onward, but the fundamental work of establishing connections through intermediary technology has only become more central to modern societies. This work, in its most evolved and influential form, takes place within the corporate organizations of the so-called tech industry. In recent years, this economic juggernaut has come under increasing pressure. Once hailed as a progressive force (Rosen 2012), the tech industry is now the subject of critical debates around exploitative business models, privacy issues, content biases, and the overarching role of digital technologies in our societies (Benjamin 2019; Blackwell 2024; Couldry and Mejias 2019; van Dijck et al. 2018; Zuboff 2019).

This context has spawned an array of investigations into the tech industry. In general, researchers have focused on the highly precarious class of gig and crowd workers, while journalists and politicians have homed in on the entrepreneurial class. However, relatively little attention has been paid to the tech workers who make up the middle layers of the industry. This is surprising given that these are the people who encode, design, and manage the digital technologies that increasingly permeate social life. Through their professional work within tech companies and start-ups, tech workers increasingly shape how we perceive the world, what we desire, and with whom we interact. However, their role and influence is largely overshadowed by the groups at the two poles of the digital class spectrum.

This overshadowing is no accident. The entrepreneurial class invests significant efforts into branding their firms' products as their individual creations. They prefer to have the world envision the latest technology as the creation of a solitary genius or a select few masterminds rather than the collective output of countless high- and low-paid workers. This practice feeds not only their egos but also, and more importantly, their business interests. Presenting their companies as technology powerhouses rather than labor-driven enterprises brings financial rewards (Irani 2015a). That is, by downplaying the labor that goes into the development of new digital technologies, tech entrepreneurs are able to attract more investment and boost the stock value of their companies.

In the field of digital labor studies (Jarrett 2022), the relative neglect of high-paid tech workers can be explained through researchers' easier access to precarious digital workers, especially true in the case of the army of gig workers who very visibly traverse city streets delivering food and goods. Another likely contributing factor is the continued investment many contemporary labor scholars have in an orthodox interpretation of Marx, in which the middle class plays no significant historical role because the potential for revolutionary consciousness resides exclusively in the proletarian class.

It must be acknowledged, however, that Marx's analysis of capitalism was constrained by his context in the nineteenth century. During his era, the middle class was so small in size that developing a binary account of class was justifiable. However, with the shift from industrial to postindustrial capitalism in the twentieth century, and the increasing relevance of knowledge and creative work, the middle-layered stratum grew significantly in Western societies. This has been recognized by an array of influential Marxists, including Wright (1976), Poulantzas (1979), and the Ehrenreichs (1979). And yet in contemporary research on digital labor and digital capitalism, the middle class is once again overlooked. Bordering on farce rather than mere tragedy, critical scholars have almost exclusively turned to the lower class and thus failed to generate a relational account of work and class in digital capitalism. Few have considered how forms of critical consciousness and the potential for social transformation could also arise from the grouping positioned in between the digital proletariat and the digital capitalists.

Against this backdrop, this book set out to inquire into the social codes of tech workers. It asks what forms of subjectivity are dominant among contemporary digital professionals and how these translate into boundary makings. In other words, the core aim was to reconstruct what makes this middle class at the professional backstage of digital capitalism *tick*, what kind of identities they cultivate, and examine how this connects to *power dynamics*. Chapters 2–4 center on interviews in the US and Germany and offered a deep dive into the hearts and minds of tech workers in the Global North. The latter chapters, 5–7, then moved to a theoretical discussion of tech workers' class formation and embeddedness within larger structures, as well as a reflection on the possibility that they are embroiled in yet another spirit of capitalism. The overarching diagnosis this book has provided is that tech workers' cultivate a *contradictory class identity*. While most tech workers are hardly revolutionaries, the majority are sharp critics of contemporary capitalism. They are disillusioned with neoliberalism despite being part of one of the few rising middle-class groups in today's society. Only a minority of my interviewees fully embrace the ideal of the entrepreneurial self with its emphasis on autonomy, creativity, and market-solutions to societal problems. Notably, a number of the social codes of tech workers can also be found among other middle-class fractions. This book contributes to the literature on knowledge workers and the "new middle class" more generally. At the same time, though, the post-entrepreneurial subjectivity that this book reconstructed gains a specific meaning because of the privileged position that tech workers occupy in the class matrix. Even though tech workers are, together with bankers and consultants, one of the few rising middle-class fractions in neoliberalism, their disillusionment with the contemporary conjuncture leads them to aspire to build, and live in, a different society.

We can view tech workers as an influential class fraction *in the making*, with the potential to tip the power balance in the tech industry toward labor. Their boundary-makings vis-à-vis market-oriented forms of subjectivity and technosolutionist ideologies hints at the possibility of reconfiguring digital technologies to align with the greater social good. For those seeking to drive social change, tech workers could play a pivotal role in forming progressive alliances. However, my study also cautions that the emancipatory potential of tech workers is far from straightforward. Tech workers' social codes are ambivalent, double-edged, and, based on my interviewees' reports, seem to rarely translate into corresponding practices. Tech workers remain entangled with deep-seated individualist principles of neoliberalism, and some of their moral frameworks have been appropriated and co-opted by tech firms into a rejuvenated moral spirit of capitalism. The story of tech workers is a story of a contradictory class identity, a testament to the ways in which capitalism and its resistances are not locked in opposition but intertwined in complex ways.

POLITICAL IMPLICATIONS

So what are the concrete political implications of the outlined identity formation? Put bluntly: Why should we care about tech workers' contradictory social codes? With regard to union-building and the possibility to democratize tech firms from within, my study demonstrates that organizing tech workers on a large scale will not succeed without a fundamental understanding of what drives them. Despite momentum in unionization efforts in the last couple of years, the vast majority of tech workers remain unorganized. This lack of collective power not only limits their ability to negotiate for better wages or to challenge layoffs, but also excludes them from having a meaningful say in the development of digital technologies. Of course, union-building is not the only avenue for collective action, but it is a particularly important way of providing workers with an *institutional base* to pursue their class interests and hold tech companies accountable from within. In their theoretical assessment of how to build "tech worker power," Boag et al. (2022) stress the importance of translating moments of dissatisfaction and subsequent employee action into collective organizations that can sustain long-term campaigns. My study tells us that tech workers are, perhaps surprisingly, sympathetic to joining these "legacy institutions," with most interviewees I spoke with affirming that they are open to the idea of joining a union. This theoretical interest in collective action also manifests in their social critique as well as in their preferred self-classification as "workers." And yet at the same time, my interviews suggest that tech workers' openness to unions rarely translates into collective action (as discussed, three-quarters were open to the idea of union membership, while only two had joined one one). Though there is evidence of "goodwill" among tech workers, this seems to be currently channeled more into moral self-monitoring than collective institution-building or the forging of new coalitions. Often, tech workers told me they could not imagine what a union at their workplace would look like in practice. These insights should motivate unions to address tech workers in new ways. Unions will need to convince these employees that they can better pursue their interests and morals if they ask what can *we* do rather than what can *I* do. Furthermore, unions will need to become more flexible, transnational, and digital, given that tech workers frequently work remotely, generally switch their employer every three years or so, and often cross national borders throughout their careers. Unions have to better adapt to this rhythm of life if they want to remain relevant.

My study also indicates the need to move beyond moral repertoires when attempting to inspire collective action. Moral aspirations for change are a necessary, but not a sufficient, condition for social transformation. What is often missing in current efforts to organize tech workers is an attempt to show them that there is *a real chance of success and tangible economic benefits.* Given increasing union-busting and employer retaliation on one side of the equation and lay-offs, the threat of AI automation and market instability on the other, mobilizing only around moral values will not be enough. For ordinary tech workers to risk their jobs and potentially even their careers, they must believe that victory is attainable and be able to visualize the payoffs of their actions. Union organizers and activists will need to pitch a winning strategy just as convincingly as tech entrepreneurs market their digital products as tools that improve lives.

One strategic path forward lies in the untapped *infrastructural power* of tech workers. Thus far, newly formed unions in the tech industry have rarely attempted to disrupt the production process (Niebler 2023b), but shifting gears toward this kind of action could prove vital to increase their bargaining power. By centering action on strategic strikes rather than symbolic acts like protest letters, unions could demonstrate the capacity of collectively organized tech workers to disrupt global supply chains and challenge the smooth flow of information in digital infrastructures.

This study also holds implications for union-building strategies within the tech industry as a whole. Given the unconsolidated state of tech workers' class formation, my research indicates that ambitions to organize the entire workforce of the tech industry under one broad conceptual banner is an unrealistic and likely counterproductive approach. This critique extends to the Tech Worker Coalition, which is pursuing initiatives that aim to unite different occupational groups by describing all who labor within the tech industry—including gig workers, cafeteria workers, and cleaning personnel—as "tech workers" (Tech Worker Coalition 2021). The organization's "Bill of Rights," which presents its organizational self-understanding, reserves no specific terminology for high-paid and professionalized workers in the digital economy (Tech Worker Coalition 2023). Activist scholars and scholarly activists have also begun flirting or advocating for a broadening of the term "tech worker" (Nedzhvetskaya and Tan 2024a; Tarnoff and Weigel 2020). While I am politically supportive of the intention of uniting different classes of digital workers, my study suggests that the conceptual strategy of broadening the notion of "tech worker" comes with too many downsides and may in fact backfire. For a start, it runs counter to the self-classifications in the industry. Most actors that I met in tech reserve the classification of "tech worker" for the high-paid white-collar ranks of workers who engineer and design digital technologies. This usage of the term also surfaces in statements by low-paid, blue-collar workers in the industry (see Tarnoff and Weigel 2020, 62–63). Moreover, expanding the concept of "tech worker" risks obscuring the distinct power positions held by different groups of workers in the tech industry. Referring to all digital workers as tech workers leaves us with no specific term with which to recognize and address the specific segment of high-paid white-collar workers in the industry. Ironically, as a consequence, this well-meaning move runs the risk of inadvertently reviving tech workers' self-classification as "professionals," which would confuse and possibly undermine their emergent and fragile class formation process.

Beyond addressing the limitations of certain strategies for mobilization, it is equally important to consider the critical role of institutional support from the political field in advancing digital labor struggles (Dencik et al. 2022). Tech workers and their unions cannot be expected to reform the industry on their own. Apart from larger initiatives, such as antitrust legislation aimed at Big Tech companies, institutional political support could include the strengthening of works councils, groups of employees who represent workers' interests in consultations and negotiations with employers. For example, in the EU, the implementation of European Works Councils has provided employees of transnational corporations with institutional leverage to pursue new types of class compromises (Franke 2023). Furthermore, political institutions could provide support by legally limiting the practice of short-term contracting, which not only affects low-paid workers but is also observed in the professional ranks of the tech industry (Tarnoff 2020), effectively establishing a two-tier system of contractors vs. employees that could impede the collective class formation of tech workers.

In terms of politics, it is important to remember that tech workers hold social relevance not only because of their potential to challenge tech firms through union-building but also because of their *inscription power*. Tech workers are directly responsible for encoding and designing the affordances of the digital technologies that increasingly permeate social life. They shape the design and functionalities of the apps we use and develop the algorithms that determine the content we engage with (Doyuran 2023). If we understand technological development as an inherently social process (Noble 1984; Ramtin 1991), tech workers, with their particular social codes, can be considered the *switchmen and -women* of our digital era. They engineer the tracks on which we traverse the virtual world. Given this highly influential position, securing their support will be crucial for many kinds of political undertaking. Political projects aiming to tackle bias (Brock 2015; Chun 2021), develop more convivial technological tools (Christiaens 2022;

Illich 1973), or build (new) forms of "platform cooperativism" (Grohmann 2023; Scholz 2016) will remain utopian dreams without the buy-in of tech workers.

Current developments around AI lend tech workers even greater powers to influence societies, for better or worse. Natural language processing and machine learning have advanced to a state where tech workers may come to be the only cohort with the expertise to genuinely grasp their operations, meaning that proactively identifying potential harms and unintentional consequences of new digital applications could hinge on the vigilance of tech workers. Now, we are seeing advancements in AI that could lead to a division among tech workers into expert developers and master deployers of AI technologies on the one hand, and a growing segment of de-skilled tech workers, who may see their upper-middle-class salaries threatened, on the other. Either way, though, educational institutions need to reconsider how they teach and train tech workers. Given the ever-greater technological influence on societies wielded by some of their graduates, it is imperative that they include more nontechnical courses in curricula so as to foster a broader perspective on the social, ethical, and cultural implications of digital tech. My analysis of study programs revealed a limited nontechnical training. While some institutions have introduced ethics courses into their tech programs, these efforts fall short of fully preparing future tech workers to engage in the socially responsible development of new technologies. A more comprehensive and holistic education is needed. Furthermore, educational institutions need to evaluate how they can address the de-skilling of some tech workers as a result of AI automation. Equipping graduates with new skills will be crucial to maintaining their employability and ensuring they are not left behind in the changing labor market.

Another institutional avenue that could be pursued to support tech workers' labor market positioning as well as their aspiration to be agents of change lies in their *formal professionalization*. To date, there exist no major associations or institutionalized bodies where data scientists, UX designers, and other tech work occupations can come together to regulate and classify their standards and values as a profession. This aligns with tech's reputation as a space that keeps its distance from established institutions and bureaucracies, but it comes with hidden costs. Professional associations play key roles in offering certification and credentials, determining ethical standards, providing opportunities for professional development, and advocating for the profession in wider political, civic, and economic contexts (Abbott 1988; Freidson 1988; Durkheim 1997). It is unthinkable, in contemporary society, that anybody could claim to be a doctor and immediately start operating on patients. Yet, in the tech industry, a scenario along these lines is possible. There is no degree, no specialized training, and no ethical oath necessary to call oneself a data scientist and start programming, unsupervised, algorithms that strongly impact people's life chances. Given tech workers' expressed concerns about the reputation of their craft, calls for increased regulation appear warranted, in their interest as well as the public's. It may be time for tech workers to relinquish their skepticism and tap into the value of old-school professionalization.

CHALLENGING NEOLIBERALISM

When considering the political implications of a set of codes, it is important to consider how far they travel. Along these lines, tech workers' codes reveal the potential for a broader movement against neoliberalism. As I discussed, tech workers' post-entrepreneurial subjectivity has been catalyzed by their experience of three major crises: (1) the economic crisis of neoliberalism, (2) the political crisis of legitimate representation, and (3) the ecological crisis of climate change. These crises also have had effects on other classes, and so, while tech workers hold a certain class location and cultivate a specific kind of post-entrepreneurial subjectivity that distinguishes them from other rising middle-class fractions such as bankers or consultants, their social codes are not entirely exceptional. A number of their cultural, moral, and political positionings can be found among other class fractions (Beck and Westheuser 2022; Grasmeier and Beck 2023; Meghji 2019; Piketty 2021). For readers who are embedded within the academic field, for instance, concerns around economic inequality and a lack of diversity combined with lifestyles of comfortable exploring will surely seem familiar. Further empirical research is required to investigate to what extent, and in what specific ways, market-oriented forms of subjectivity may be running out of steam across the middle class and beyond. What is already clear is that unrest with neoliberal capitalism is visible among different status segments of contemporary society, as reflected in recent strike actions by teachers, graduate students, creative workers, or gig workers. There is, perhaps, something in the air, something up for grabs. We appear to be living through a *hinge moment* in capitalism, where, how Gramsci once put it: "the old is dying but the new cannot yet be born" (cited from Streeck 2016, 69). This study signals how a profound dissatisfaction with the status quo can be found even among one of the few rising class fractions in society, one that does not feel the economic squeeze like most other middle- or lower-class groups but remains fundamentally disillusioned with the contemporary conjuncture nonetheless. This indicates a certain potential for building powerful class coalitions across status segments. Without such broad alliances, seizing this pivotal hinge moment and preventing "monsters" from taking over will be an uphill battle.

However, building such coalitions will require bridging the cultural divides between liberal and cosmopolitan identities, on the one hand, and more communitarian and rooted identities on the other. This study of tech workers tells us that increased self-reflection will be necessary, including, for instance, that a lifestyle of comfortable exploring may exclude certain social groups without the resources to keep up. Such awareness could help to avoid latent devalorizations of the lifestyles of the less affluent. Furthermore, critiques focused on a lack of diversity will need to avoid succumbing to capitalist elites' superficial appropriation of the issue. Cultivating broad class coalitions will also require actors to focus more on modes of existence that are widely shared (Lamont 2023). This could happen through a shared appreciation of labor, ordinariness, and a concern for social justice that operates without casting others as inherently less worthy.

Building class coalitions must also happen at a global scale. Overcoming the culture wars in Western societies and increasing worker power locally, as challenging as that may prove, will not be enough to seriously confront the power of capital and take advantage of the conjunctural hinge moment we are living through. The transnational nature of capital, which is particularly advanced in the tech industry, cannot be tackled by locally organized labor movements alone. Notably, one profound and often overlooked dialectical insight in Marxist theory is that capitalism brings about the very conditions and tools required to challenge it (Hardt and Negri 2000). For Marx, factories were alienating spaces, but they also brought members of the dominated class closer together. The shift from feudalism to capitalism threw peasants into a previously unknown physical proximity with each other as industrial wage laborers. This generated various hardships but also served as a fertile ground for members of the dominated class to forge interpersonal bonds, exercise shared cultural repertoires, orchestrate workplace resistance, build unions, and form new political parties. In contemporary digital capitalism, labor's chances of challenging capital's dominance seem to have diminished. Digitalization seems to pull us apart, atomizing us in front of our screens, propelling heightened forms of surveillance, and intensifying competition among workers (Fourcade and Healy 2024; Graham et al. 2017). While acknowledging this, we must also consider the chances that digitalization brings with it. When contemplating about more global class alliances, we need to think carefully about the tools and spaces that can make it happen. Without wishing to advocate for a technosolutionist agenda in the final pages of this book, I hold that tech workers' know-how must be considered vital to any such effort. If deployed in a critical and reflective manner, their expertise brings with it the possibility of building new technical tools and platforms from which to challenge capital. We have already seen an illustration of this in the case of transnational cooperation between Chinese and US tech professionals (Tan and Weigel 2022), where US-American tech workers' insistence on keeping GitHub open and uncensored allowed Chinese tech workers to continue their labor activism using the platform. We can also think of platform cooperatives and the development of platforms that operate outside the profit motive, such as Wikipedia or Mastodon, as examples, despite all their flaws, where digital expertise has allowed for processes of decommodification and brought people together in virtual spaces that are not run by and for the market.

POST-ENTREPRENEURIAL SUBJECTIVITY AND SOCIAL THEORY

Embarking on our journey into the grouping of tech workers and their social codes began with a consideration of the theoretical literature that has investigated the nexus of subjectivation and capitalism. That is, we conducted a deep dive into the hearts and minds of tech workers, but not without putting on our theoretical goggles first. From the start, this study differentiated two theoretically distinct and much-studied forms of subjectivity and corresponding modes of capitalist production: the organizational self, which describes the ideal-typical subjectivity of white-collar workers in Fordist capitalism, and the entrepreneurial self, which describes a more market-oriented social figure that gained dominance with the rise of post-Fordism and attendant economic shifts toward deregulation and flexibility evident from the 1970s onward. One central theoretical contribution of this book has been to demonstrate that the new middle-class fraction of tech workers operates differently. They not only deviate from but actively distance themselves from the figure of the entrepreneurial self. Without positing a complete break with neoliberal ideals, or a simple reversion to the organizational self, this study has revealed the contours of a *post-entrepreneurial subjectivity* within contemporary capitalism's leading industry. Importantly, this conceptual argument should prompt us to critically examine how tech workers, and other middle-class fractions who share some of their codes, will respond to recent shifts at the political-economic level. The 2024 reelection of Donald Trump—supported by a number of tech leaders—may structurally alter the current conjuncture in a number of ways. Already, it has gone hand-in-hand with a cultural shift within the entrepreneurial-managerial class, which appears to be increasingly moving away from co-opting liberal social critiques as part of their strategies. Grasping whether tech workers' partially realized, contradictory class identity will be transformed in this context requires further social scientific investigation. Tech workers are a class in the making, and their post-entrepreneurial subjectivity is a freshly carved identity.

Finally, there is another, more abstract social-theoretical move that has lurked somewhat in the background thus far: a sociological program that sees emancipation and domination within the same frame. The transformation of the entrepreneurial self cultivated by tech workers opens up a space of possibilities in both directions of power. Building on this insight, I posit that we would fail to grasp the logics of most, possibly even all, social phenomena if we focus exclusively on either the dimension of domination or of emancipation. It is worth noting that Lamont has followed up her seminal studies on boundaries with explorations of bridge-makings (Lamont and Aksartova 2002) and of resilience and recognition (Hall and Lamont 2013; Lamont 2023). Many scholars and theorists of work and culture in the digital economy have consciously departed from a paranoid mode of critical analysis fixated only on revealing domination (e.g., Amrute 2016; Irani 2019; McPherson 2023; Su et al. 2021). Still, Bourdieu and Foucault's more abstract social theoretical frameworks, which have been fundamental to this study, are primarily concerned with processes of subordination and the complicity of ostensibly progressive institutions of modernity in economies of control and class reproduction. While both Bourdieu and Foucault showed an interest in forms of resistance to relations of domination in their later works (Bourdieu 2003; Bourdieu 2008; Foucault 1988, 2012), their oeuvres are mostly known for bringing to light the increasingly latent and subtle forms through which social inequalities are (re)produced. This legacy has meant that their theoretical frameworks are often deployed in ways that neglect emancipatory processes and potentials (Rosa 2019). I hold that we need to rectify this oversight, attending to emancipation and resistance while preserving the importance of exploring domination through latent strategies and the micro-physics of power. While it remains an empirical question to what extent and in what forms processes of domination and emancipation manifest in certain social relations, our theoretical heuristics should be open and sensitive to both dimensions. The space of possibilities inherent to every social field, to every discursive formation, and to every subject formation, must be evaluated for its capacity to contribute to both the stabilization and destabilization of the class system and power-knowledge nexus. As this study has shown, even switchmen and -women ride on multiple and colliding tracks.

Acknowledgments

From a sociological perspective, the notion of subjectivity never invokes a fully autonomous actor; rather, human beings achieve subjectivity through their embeddedness in social relations. Consequently, I could not have authored this work without the support of an array of people and institutions. First of all, I wish to thank Jennifer Gabrys for her encouraging support and wise guidance in the years during which this research came about. Her mentoring has taught me so many valuable things about the practice of research. I also want to thank Andreas Reckwitz for many fruitful discussions about how my object of analysis relates to larger sociological questions. I am very grateful to Justin Kehoe and Winifred Poster for their invaluable support as editors of the MIT Press Labor and Technology book series. Thanks are also due to Katherine Dhurandhar, Jitendra Kumar, and Roger Wood at MIT Press for the copyediting. Furthermore, I am indebted to Steffen Mau and Philipp Staab, who have continued to meet with me and invite me to their colloquia even though my time at Humboldt-University Berlin has long ended. In terms of academic support, inspiration, and invaluable feedback on my work, I also need to thank Jutta Allmendinger, Philipp Brandt, Tim Christiaens, Roisin Curtin, Iman Delimustafić, Mathis Ebbinghaus, David Ewing, Sam Friedman, Jan Fuhse, Valentin Ihßen, Jannis Koltermann, Lilian Kroth, Valentin Niebler, Ella McPherson, Gillian Moore, Mark Ramsden, Sebastian Raza, Matthew Sparkes, and Jan Wetzel. There are many other individuals who have shaped my path and to whom I am grateful. I also owe thanks to a number of institutions whose force fields I was allowed to enter. I would especially like to thank the Department of Sociology at the University of Cambridge, the Weizenbaum Institute in Berlin, the LaborTech Research Network, and the Foundation of the Germany Economy (Stiftung der deutschen Wirtschaft) for providing me with the academic and economic resources necessary to pursue my project.

In closing, I wish to turn briefly to the private realm. This work would not have been possible without the support of my parents, Martin and Susanne. Among many things, both taught me the value of critically questioning taken-for-granted assumptions about the world and of making up one's own mind as far as that is possible.

* * *

Some parts of this book have been published in the following peer-reviewed, open-access journal articles:

Dorschel, Robert. 2021. "Discovering Needs for Digital Capitalism: The Hybrid Profession of Data Science." *Big Data & Society* 8 (2): 1–13.

Dorschel, Robert. 2022a. "A New Middle-Class Fraction with a Distinct Subjectivity: Tech Workers and the Transformation of the Entrepreneurial Self." *The Sociological Review* 70 (6): 1302–20.

Dorschel, Robert. 2022b. "Reconsidering Digital Labour: Bringing Tech Workers into the Debate." *New Technology, Work and Employment* 37 (2): 288–307.

Dorschel, Robert. 2024. "Middle-Class Responses to Climate Change: An Analysis of the Ecological Habitus of Tech Workers." *Current Sociology* 72 (7): 1301–18.

In the book, the articles are not reproduced in full. Rather, the book builds and expands on the articles across all chapters. Nevertheless, some chapters are more closely related to certain articles than others. Chapters 2, 4, and 6 draw closely on the article in *The Sociological Review*; chapter 1 builds on the piece in *New Technology, Work and Employment*; chapter 3 draws closely on the article in *Big Data & Society*; and chapter 4 builds on the piece featured in *Current Sociology*.

Appendix: Analytical Strategy

This appendix will serve to expand and elaborate on the discussion of methodology in the introduction. In the first section, I will discuss my analytical strategy by providing more information on my research design, including the choice of interviews and discourse analysis as methods of my inquiry. In the following section, I will elaborate on the deployment of these two methods by discussing in detail the processes of data collection and data analysis. In the third section, I will reflect on my positionality and highlight central limitations of my research design in terms of methods and data. Across these sections, it will thus become clearer how I approached (and thus also constructed) my object of analysis as well as the procedures I undertook to reconstruct the social codes of tech workers.

My analytical strategy is built around a mixed qualitative methods approach using both interviews and discourse analysis. By combining these two methods, my research design follows the program of twofold subjectivation studies (Bosančić et al. 2021; Pfahl et al. 2018; Spies 2017), which posit two distinct methodological entry-points into processes of subjectivation: one that focuses on the meaning-making processes of subjects themselves, and one that contrasts these with ordering processes at a more macro-sociological level. This dual methodological perspective corresponds with abstract theoretical frameworks of subjectivation and habitualization in the sense that actors are understood to navigate a delimited space of possibilities that arises from their embeddedness in social structures (Bourdieu 1990; Foucault 1982).

At the level of subjective meaning-making processes, interviews provide a crucial gateway into understandings and presentations of the self. This method allows me to grasp how tech workers make sense of themselves and the social world in which they are embedded. While debate remains ongoing as to the relation between how people present themselves in interviews and what they actually do (Khan and Jerolmack 2013; Maxwell 2012; Vaisey 2009), the method can still be considered the most suitable approach for unearthing "where people live imaginatively—morally but also in terms of their sense of identity—what allows them to experience themselves as good, valuable, worthwhile people" (Lamont and Swidler 2014, 159). Discourse analysis, on the other hand, allows me to gain insights into the interpellation of subjects through institutionalized knowledge orders. Following Foucault, discourses are understood as systems of utterances with relatively autonomous formation rules (Foucault 2002). A discourse exerts subjectivation effects by providing identity offerings (Keller 2011). Discourse analysis was mobilized to provide an account of larger sociocultural ordering processes in the context of which tech workers make sense of themselves and their social world.

I also hold that the combination of interview and discourse analysis is particularly well-suited to the study of emergent social phenomena. The relative newness of tech workers' occupational rise and differentiation means it has not yet been comprehensively picked up by formal surveys and typologies administered by institutions, including the US Bureau of Labor Statistics or the SOEP panel in Germany. In such contexts of emergence, the analysis of subjective and discursive classification systems constitute vital entry-points for understanding the ongoing institutionalization process of social groups (Goldstein 1984; Lamont and Swidler 2014). Thus, while further methods would yield additional insights, I maintain that the two selected methods are sufficient for a thick explorative account of the subjective and discursive classifications of tech workers.

DATA COLLECTION AND ANALYSIS

My data collection and analysis is guided by the basic principles of grounded theory (Glaser and Strauss 1967) and abductive research (Timmermans and

Tavory 2012). Both programs are oriented toward developing new theories through a back-and-forth between data collection, data interpretation, and theory building. However, abductive research departs from grounded theory in the sense that it operates less inductively, leaving more space for the inclusion of theoretical heuristics from the onset of the empirical research. This methodological procedure very much aligns with the basic aim of this work, which draws on social theory from the beginning in its approach to tech workers, and interprets their social codes against the backdrop of theoretical conceptualizations of middle-class identity through history.

In the data collection phase of interviews, I drew on Bourdieu's field theory (1996) to coordinate the sampling process. Given the vast number of professionals who render digital technologies at internet-related companies and earn upper-middle-class wages, I focused on two groups of tech workers with distinct positions in the professional field: data scientists and UX designers. I interviewed fifty-two data scientists and UX designers in three waves, virtually via Zoom, between July 2020 and March 2022. The total number of interviews fell short of offering an exhaustive account of tech workers' subjectivity, but it did allow for sufficient data to address my research interests in theoretically saturated ways. That is to say, the number of interviews was enough to identify recurring patterns in the self-understandings and -presentations of tech workers that demonstrate a contradictory transformation of the entrepreneurial self.

My exploratory sample comprises thirteen data scientists and thirteen UX designers in each of two countries: Germany and the US. As discussed in the introduction, the two national field sites were chosen due to their economic relevance and because they form a contrastive pair in the Global North: Germany can be considered a "coordinated market economy" with a relatively strong welfare state, while the United States is regarded as the ideal-typical "liberal market economy" with generally lower taxes and lesser social services (Hall and Soskice 2001). While both economies have undergone neoliberalization, it is worth considering that the process in Germany dates from the early 2000s, where it was implemented by a center-left coalition, while in the US the deregulation of the economy and the scaling back of the welfare state started under the Reagan administration in the 1980s and has enfolded in a starker fashion. The two nation-states are also relatively contrastive in the sense that Germany has retained a comparatively strong industrial sector and corresponding workforce, whereas the US economy relies more heavily on finance, media, and tech. Given these differences, I hold that identifying similarities between actors across these two settings makes it possible to advance, albeit tentatively, a general account of tech workers in the postindustrial Global North.[1]

Most of my interviewees work and live in either the San Francisco Bay Area or Berlin—the two biggest tech hubs in the respective countries. However, I also interviewed fourteen tech workers from a few other cities in the US and Germany to reflect on city-specific phenomena. As outlined in the introduction (where I also discuss the class, gender, and racial backgrounds of my interlocutors), the mean age in my interview sample was thirty-four, with the ages of my interviewees ranging from twenty-five to fifty-three. Thus, my findings may be influenced by generational effects in the sense that a lot of tech workers belong to the "Generation Y." At the same time, however, my findings contrast with recent studies of other upper-middle-class professional segments where many members of the Generation Y are drawn to, such as finance (Hofstätter et al. 2016; Neckel et al. 2018; Neely 2022).

I deployed multiple strategies in the recruitment of interview partners to minimize possible selection biases. I recruited participants through personal contacts, for which I relied on the contacts of friends, colleagues, and professors whom I had met as a graduate and postgraduate student in Berlin, Durham (US), and Cambridge (UK). I believe this allowed me to reach individuals who would have been reluctant to participate without a common friend or contact. Deploying the snowball technique, I then asked my interviewees to provide me with further contacts. Furthermore, I was able to recruit interview partners outside my personal network through the online platform LinkedIn. This technique of "cold calling," or rather "cold contacting," worked surprisingly well—perhaps due to the symbolic capital that being a PhD student at the University of Cambridge granted me. Last but not least, I recruited participants through a membership at a coworking space in Berlin that I joined during the Covid-19 pandemic. This coworking space is located in the center of Berlin and is one of the biggest in the city. In addition to contacting people there directly, the coworking space also offered an online directory that allowed me to message tech workers who had moved to the United States.

I began the semi-structured interviews with an open question about how participants became involved in tech work. Starting with an open question was intended to signal to interviewees that they may speak freely and tell me their story in depth. After this, I would ask questions concerning their worldviews, for instance, about what bothers them at work and in society. I then asked questions related to their everyday work, for instance, what they do on a daily basis and what characterizes a good tech worker in their opinion. I concluded the interview with questions about their personal lives and social backgrounds: For example, I would ask how they spend their leisure time and what their parents do/did for a living. The virtual interviews lasted between forty-five and ninety minutes, and on average a little more than one hour. Before the interview, all participants signed a consent form where they agreed to be interviewed and to be recorded. Furthermore, I introduced myself and told them about my basic research interests and informed them that I was not interested in acquiring specific information about the development of current technologies. This latter information usually allowed for ease, enabling my interviewees to open up more freely. Furthermore, I aimed to foster a productive interview setting by always showing interest in what interviewees told me, while seeking to avoid sending positive or negative feedback, let alone appearing opinionated or judgmental. Nonetheless, I cannot rule out that a certain interviewer effect took place in the sense that some of my interviewees may have felt reluctant to express certain values if they assumed that a university researcher would not share these. The interviews were saved on an external and encrypted hard drive. Transcripts were always anonymized despite two research participants expressing their willingness to be quoted by name.

In addition to conducting interviews, I mobilized discourse analysis to examine the institutional construction and interpellation of tech workers' subjectivity. This method is typically applied through the analysis and interpretation of textual documents (Foucault 2002; Keller 2004, 2011). The textual corpus of my discourse analysis (*n* = 188 documents) is composed of data science and UX design study programs as well as job ads published between March 2019 and June 2021. While the discursivations within study programs and job ads cannot be taken at face value as representations of nondiscursive processes, both mediums can be considered obligatory passing points for the development of tech workers. In some way, tech workers must position themselves vis-à-vis the expectations raised in job ads and study programs. These textual materials are highly visible and generate insights into how firms and universities imagine ideal graduates and elite professionals in the field (Ingram and Allen 2019).

The documents of my textual corpus were gathered via the homepages of university websites. For the academic field, in which, according to Bourdieu, the most important actors are the universities (Bourdieu 1988, 128), I collected materials related to data science and UX design study programs ($n = 50$) from high-ranked, medium-ranked, and low-ranked universities in both the US and Germany (see tables A.1 and A.2).[2] While I was able to

Table A.1
Data science master study programs

US	Germany
Boston University	Free University of Berlin
Brown University	Leuphana University
Columbia University	LMU Munich
DePaul University	RWTH Aachen
Duke University	TU Braunschweig
Georgetown University	TU Chemnitz
Harvard University	TU Dortmund
Illinois Tech Institute	TU Munich
Northeastern University	University of Bielefeld
Oklahoma University	University of Jena
San Francisco University	University of Leipzig
Southern Methodist University	University of Mannheim
Stanford University	University of Marburg
University of Virginia	University of Potsdam
Virginia Data Science School	University of Saarland

Table A.2
UX design master study programs

US	Germany
Baltimore University	Hochschule Magdeburg
Carnegie Mellon	LMU Munich
De Paul University	RFH Cologne
Indiana University	Ruhr West Hochschule
Jefferson University	TH Ingolstadt
Kennesaw State University	University of Hamburg
Santa Cruz University	University of Potsdam
University of Georgia	University of Siegen
University of Maryland	University of Weimar
Washington University	University of Würzburg

collect fifteen data science study programs in each country, I collected only ten UX design or UX-related study programs due to their limited availability in Germany. I selected universities that included detailed descriptions of the aims and goals of the study programs as well as information on their curricula. For the economic field, in which, again according to Bourdieu, the central actors are firms (Bourdieu 2005, 75), I collected job ads ($n = 138$) from the popular online platform Glassdoor. I selected job ads that included a detailed description of both the necessary qualifications expected of applicants and the responsibilities of the role. I sampled job ads from established and start-up tech firms primarily in Berlin and the Bay Area (see tables A.3 and A.4). Only digital companies, also referred to as tech firms, understood as companies with internet-based business models, were used (and not, for instance, companies in other industries, such as traditional automobile firms).

Data Analysis

Regarding the phase of data analysis, my research design loosely follows the guidelines set out by Maxwell (2012) as well as Keller (2011, 2013). Following Maxwell, I jotted down notes for ad hoc memos during the

Table A.3
Data science jobs ads (number of analyzed materials in parentheses)

US	Germany
Established tech firms:	Established tech firms:
Amazon (4)	EVENTIM (4)
Apple (4)	Zalando (4)
Google (4)	WOOGA (2)
Microsoft (4)	Scout24 (2)
	Delivery Hero (2)
	HelloFresh (2)
Start-up tech firms:	Start-up tech firms:
JUUL Labs (3)	TIER (2)
WeWork (3)	Deevio (2)
InstaCart (3)	ShareNow (2)
Stripe (3)	Eyeo (1)
DoorDash (3)	Zero to One Search (1)
Slack (2)	Remerge (1)
Discord (2)	Smart Steel (1)
Metromile (1)	Fliit Holding (1)
Blue Owl (1)	Omio (1)
	Regis 24 (1)
	Körber (1)
	Hundred 5.0 (1)
	Marley Spoon (1)
	Zattoo (1)
	Ultimate AI (1)
	Frontier Car Group (1)
	ORAYLIS (1)
	Searchtalent (1)

Table A.4
UX design jobs ads (number of analyzed materials in parentheses)

US	Germany
Established tech firms:	Established tech firms:
Amazon (4)	Aperto (4)
Facebook (4)	Delivery Hero (4)
Google (4)	ImmoWelt (2)
Microsoft (4)	SAP (2)
	Zalando (4)
Start-up tech firms:	Start-up tech firms:
DoorDash (3)	Advocado (2)
Coinbase (2)	Cozero (1)
Datatron (2)	Enersis (1)
Eluvio (1)	Healy World (1)
Envoy (1)	Marley Spoon (1)
Grail (1)	ONE LOGIC (1)
Instacart (1)	Project A Ventures (1)
Ongo (1)	SevDesk (1)
Padlet (1)	SnapAddy (1)
PayJoy(1)	Tado (1)
Rumble (1)	TaxFix (1)
The Yes Product Designer (1)	Visible (1)
	Workpath (1)
	Xentral ERP (1)
	Yazio (1)

interviews. After the interviews were completed, I immediately turned to transcribing the recordings. In a first step, this was done with the assistance of the automatic transcription software otter.ai. Like in many other cases, the "AI" component here still requires human input to do the work successfully. In other words, it was necessary to manually check and edit the automatically generated transcriptions. After this step was completed, I would then upload the finalized transcriptions into the qualitative text analysis software MAXQDA. I did the same with the textual documents that were collected for the discourse analysis. In separate files and following established methodological guidelines, the documents were then read and coded with regard to themes and issues (Keller 2013, 114; Maxwell 2012, 116). Memos were used to capture initial theoretical thoughts, narrative elements, and general impressions. In a second step of the data analysis, the texts were re-read several times and more interpretative, as well as more abstract, codes were applied. The following extract from an interview with Michael, a data scientist from the US, serves as an illustrative example for these steps of data analysis:

> There are a lot of clever words about barriers to entry and network effects and things like that. But let's call it what it is there. There's nothing a venture capitalist wants to hear more in a pitch meeting than how you're going to be a monopoly. And that's I think a morally and ethically bankrupt position to take.

In the first part of the data analysis, this utterance was attributed the two more descriptive thematic codes of "critique of monopolies" and "venture capitalists as morally bankrupt." In the later stage of data analysis, both codes contributed to the more abstract and theoretical social codes termed "return of social critique" and "distinction vs. upper classes." Furthermore, the above interview extract also served as reference for a memo concerning the theoretical implications of "moral value systems."

As for the presentation of my data analysis, chapters 2, 3, and 4 of this work present empirical findings that correspond to the more abstract social codes that structure typical processes of subjectivation in the field of tech work. The presentation of social codes is accompanied by a select number of interview extracts. Throughout the empirical chapters of this study, my aim is to privilege analysis over description and to generate theoretical insights.

POSITIONALITY, LIMITATIONS, AND AVENUES FOR FUTURE RESEARCH

I first became interested in the social phenomenon of tech work when I attended a university course in the newly introduced study program "Interdisciplinary Data Science" at Duke University as a visiting master's student in Fall 2018. I had originally joined the course to learn quantitative text analysis, when I found myself in a room full of people who, it seemed to me, were not just sharing a master's program but embarking on a common journey. They also kindly enlightened me about the promising career opportunities offered by the tech industry. My interest in this grouping grew as many of my peers and friends, including numerous sociologists, decided to apply for jobs at tech firms and start-ups. Reflecting on my positionality, I can thus say that my academic training and middle-class position brought me into contact with the new occupational segment of tech workers. Importantly, these two factors would also influence my research process in many ways. For instance, at the coworking space in Berlin, in which I recruited a number of interview partners, I blended in very well. The fact that I am white and male only added to my inconspicuousness. Tech workers often read me as an insider at first sight. It was only once I introduced myself as a sociologist and elaborated on my research interest that I was read as an untypical figure in the field.

Throughout my research, I continuously reflected on how my own subjectivity influenced the way I navigated the field and reached my findings. Such reflection was especially crucial during the interviews, where I was aware that my presence in a data generating situation influenced the data I would retrieve. Here, it must be noted that I conducted interviews exclusively online via Zoom. This prevented me from accessing certain insights I might have expected from a live interview, such as embodied interactions and visiting my interviewees in their homes or at their offices. At the same time, I noticed that all my interviewees were very experienced and appeared comfortable using Zoom. It seemed to me that invitations for Zoom interviews had a low threshold for participation and that they may have reduced the interviewer effect in the sense that my habitus was less present due to the virtual setting of the interviews.

It is also significant that I started my research in 2020 just prior to the US presidential election contested by Donald Trump and Joe Biden. Thus, the first wave of interviews especially took place in a context of political polarization, which may have influenced my interview partners. It's worth noting, though, that my findings about US tech workers closely resemble my findings about tech workers in Germany—a national setting that is considered less politically polarized (Boxell et al. 2024; Westheuser 2022). Another limitation of my research is its focus on tech workers in the Global North. While I did interview a number of tech workers in the US and Germany who had migrated from the Global South (and I offer a preliminary analysis of their distinct experiences, especially in chapters 3, 5, and 7), my research is ultimately focused on the situation within two leading national economies of the Global North. There is thus a need to expand on this study by analyzing tech workers in other settings. Studying tech workers in the context of digital hubs outside the Global North promises to yield important insights that could inform debates about varieties of digital capitalism; the (re)production of social inequalities, seen from a global perspective; and the possibility of transnational class making processes. This line of inquiry could also take account of the experiences of tech workers who migrate back to the Global South after having worked in tech hubs such as Berlin or San Francisco.

Relatedly, future research building on this work could also be pursued through a more general study of subjectivation processes in the middle layers. As noted in the previous chapter, this study raises questions as to whether a more widespread moral shift of social character is underway within the middle class. Another central avenue for future research on tech workers lies in the application of additional methods. As I note in the introduction, a key limitation of my research is the fact that I draw exclusively on interviews and discourse analysis. Further methods are necessary to gain a truly exhaustive account of tech workers as subjects and as a class fraction. Ethnographic and quantitative analysis, in particular, could open new insights.

The ethnographic approach could allow for complementary insights into tech workers' everyday lives, including new perspectives on the (potential lack of) translation between their self-understandings and their practices. Furthermore, observing the practices of tech workers could generate rich, multifaceted accounts of their entanglements with an array of non-human actors, constituting a case study of particular relevance for science and technology studies (Callon 1984; Gabrys 2011; Latour 2005; Wajcman 2006). One especially productive avenue for future research along these lines may involve bringing symbolic boundary-makings into dialogue with the role of "boundary objects" (Bechky 2003) such as software programming tools (e.g., R or Python) or code sharing platforms (e.g., GitHub). Additionally, quantitative empirical studies could offer a more comprehensive account of tech workers' socio-structural positions and backgrounds, a necessary foundation for advancing beyond the connections I have drawn here between the socio-structural characteristics of tech workers and their cultural, moral, and political schemas. Combined with a qualitative approach, quantitative research could further contribute to a more holistic understanding of how the tech industry, often still framed as an inclusionary and meritocratic world of work, continues to operate via intersecting forms of class, gender, and race-based forms of exclusion.

Notes

PROLOGUE

1. All names of my interviewees were anonymized.

CHAPTER 1

1. Various definitions of class exist in the social sciences. My understanding of class follows a multidimensional approach by addressing different forms of capital (Bourdieu 1984; Lamont 1992). In terms of objective economic class positions, studies typically define the middle class as being located in between 65%/70% and 150%, 200%, or 250% of the median income (Atkinson and Brandolini 2013; Pressman 2015). Against this backdrop, my interviewees must be considered at the upper end of this spectrum.

2. Further studies of tech workers could be listed here: I will point here to two major examples, which I also engage with in later parts of this book. The first is a recent work by Chen (2022) that has analyzed the religious underpinnings of tech culture in Silicon Valley. Importantly, this insightful study is based on research with both tech employees and entrepreneurs, between which Chen does not systematically differentiate. I hold that this leads her to overemphasize the importance of entrepreneurial schemas among tech workers. Second, there is the recent study of civic tech organizations in Silicon Valley by Rider (2021). This work provides important insights into the members of organizations who volunteer in their free time to build technology for the "common good."

3. Di Di's (2023) as well as Tan and Weigel's (2022) comparative studies of tech workers in the US and China are important exceptions that I will engage with in chapter 5 and 7. Furthermore, Sidney Rothstein (2022) has provided an impressive historical comparative study of "tech workers" that I will discuss in more detail in chapter 5. Notably, Rothstein's study does not examine the new generation of tech workers within contemporary internet-based companies but focuses on workers who are embedded in more traditional technology firms.

4. Of course, however, we may identify several significant differences between Foucault and Bourdieu. First and foremost, while Foucault links the incorporation of dispositions and schemas primarily to discourses, dispositifs, and technologies of the self (Foucault 1982, 2002), Bourdieu is more interested in the links with different forms of capital, social classes, and social fields (Bourdieu 1984, 1996). Furthermore, in contrast with Bourdieu, Foucault focuses more on subjects' capacity to reflect and self-transform, at least in his later works (Foucault 2012a, 2012b)—this is why I prefer the notion of subjectivity over habitus. However, it's important to recall their commonalities: Both authors place great importance on the incorporation of dispositions and schemas at the individual level and choose to inquire into this dimension through a relational framework with a focus on the modes of power and domination.

5. My deployment of the notion of codes differs from the popular deployment by Alexander (e.g., 2004, 530) in the sense that I want to suggest that codes have the potential to act as both something actors can deploy (Swidler 1986) and a meaning structure that stirs the actions of actors (Vaisey 2009). Furthermore, building on Lamont, I reserve the notion of moral codes to other-oriented normative codes (Lamont 1992, 4, 184).

6. This is the reason why I link the concept of subjectivity to "codes" instead of "classifications," which is the more popular concept but does not semantically sensibilize us for the normative dimension of social life.

7. The issue of where the boundaries of the tech industry lie must thus be regarded as an empirical question. The fact that more and more long-existing companies are restructuring their businesses around the commercial internet and are adopting tech-like organizational cultures could be discussed as moments of their field entry into the tech industry.

8. I considered companies to be start-ups if they were less than ten years old and were oriented toward innovation and fast growth.

9. With regard to the upper class, Sherman (2019) has shown how these are guided by a similar hybrid aspiration.

CHAPTER 2

1. This is also true for gig workers, who weren't considered to be "tech workers" by any of my interviewees. This contrasts with the classification principles used by organizations such as "The Tech Worker Coalition," which pursues a political definition by referring to all employees working for tech companies, including gig workers as well as cafeteria or cleaning staff, as tech workers. While I am sympathetic to this political rationale, I hold that it is important to differentiate between a sociological definition and a political definition of tech workers. See Niebler (2025) for a discussion of the divergent uses of the term "tech worker".

2. This latter form of distinction can be considered to contribute to the establishment of a social field that we may call the *tech field*.

3. Aside from important exceptions, such as in matters of policing and security (Wacquant 2010).

4. This finding stands in some contrast with claims made in Chen's study of Silicon Valley work culture (2022). I interpret that the different findings are largely due to the fact that she interviewed not only tech workers but also tech entrepreneurs.
5. Notably, the Alphabet Workers Union is not (yet) formally recognized by the National Labor Relations Board.
6. The conversations by tech workers in online professional exchange communities studied by Sengupta and Tacheva (2022) also demonstrate widespread discrimination experienced by women in tech while simultaneously showing how women-centric online forums can serve as spaces for building resilience and empowerment.
7. Personalization and customization constitute a client-orientation of tech workers that can be considered a typical ethos of newer "corporate professionals" (Muzio et al. 2011, 451).

CHAPTER 3

1. I wish to underscore that by attesting social actors an *illusio* I do not conceptualize them them as free of critical and reflective capacities. While an *illusio* constitutes a deep-layered meaning and belief system, actors are at least partly aware of why and how they engage in the action and struggles of a social field (Boltanski 2011).
2. In this diagnosis, morals are understood to be linked to beliefs about whether behavior entails consideration for others as well as whether it is "good" or "bad" behavior.
3. The concept of "post-nerdiness" parallels, in a certain sense, usage of the term "post-structuralism," which is typically deployed to describe the theories of authors who have transformed rather than fully broken with structuralist thought (e.g., Foucault 1995, 2002). When speaking of "post-nerdy" as well as "post-entrepreneurial" subjects, I thus imply a transformation of, rather than a complete departure from, established forms of subjectivity.
4. In another contrast with traditional professions, no widely established professional associations exist among data scientists and UX designers.
5. The only other upper-middle-class profession where we might find this degree of heterogeneity in terms of academic backgrounds is consultancy (Schmidt-Wellenburg 2014).
6. Lahire's post-Bourdieusian work (2011) engages in depth with the question of how habituation processes play out across social fields and terrains with distinct logics.
7. It would be interesting to engage in further comparative study of rather novel tech work professions, such as data science and UX design, with more established ones, such as software programmers (see Beecham et al. [2008] and Boes et al. [2018] on the latter).
8. Given the size of start-ups, they often require tech workers to handle an even greater variety of tasks than tech workers at big tech companies must do.
9. During the interviews, tech workers rarely mentioned that they require both generalist and specialist technical skills. Only a few interviewees hinted at this when they told me that start-ups require more generalist capacities from tech workers than do big tech companies.

The subjectivation pattern of constructing tech workers as generalist and specialist is much more widespread in the discourse.

10. It is important to note that the teaching of communicative skills is more frequent in US academic institutions, which can be interpreted as constituting a competitive regional advantage with regard to human capital vis-à-vis Germany. Furthermore, the fact that higher-ranked universities place greater emphasis on teaching not only technical skills but also social skills means this can be interpreted as a strategy that allows for the reproduction of hierarchical relations between universities within the academic field.

CHAPTER 4

1. This work conceptually differentiates between "lifestyle" and "conduct of life." While the former is considered to point to tastes and leisure activities, the latter is understood to encompass the entirety of schemas of perception, recognition, and action of an actor (Müller and Sigmund 2014, 114).

2. According to Turner (2008, 241), a proto-version of the post-Fordist world of work can actually be found in the academic world. In the context of the Manhattan Project in the 1940s, the scientific discovery and creation of the nuclear bomb was paired with an institutionalization of network structures and a project-based work setting (see also August [2021] on the historic institutionalization of network structures).

3. This observation should lead us to question the sociological tendency to attribute actors the same degree of agency on an a priori. Put bluntly: it seems to me that the structure-agency dilemma must be resolved on an empirical case-by-case basis rather than through generalizing theorizing.

CHAPTER 5

1. Notably, though, different terminologies are often deployed in cultural class analysis. "Reflection," "collectivity," and "solidarity" may or may not be explicitly used.

2. Importantly, all three principles can also manifest in a class formation of individuals with the common interest to uphold the status quo or even intensify relations of domination.

3. In turn, this also means that a quarter of my interviewees (those whom I have considered under the ideal-type of "affirmative tech workers") are not systematically engaging in processes of class formation. The grouping of tech workers is thus no homogenous bloc with one sole trajectory in social space.

4. The extent to which Cambridge Analytica enabled Trump's victory, however, is controversial. My argument is not aimed at considering Donald Trump's victory as predominantly made possible by algorithms.

5. Of course, critiques of insufficient diversity overlap with critiques of economic relations. In this section, though, I want to highlight the different logics and specific ties to capital and class that are at play.

CHAPTER 7

1. Notably, there have been important criticisms of the various accounts that theorize the middle class across the Global North as in a state of heightened polarization (e.g., Beck and Westheuser 2022; Kumkar and Schimank 2021; Mau et al. 2020). Due to space, I cannot recapitulate this debate in detail here. However, I will show how the subjectivity of tech workers reveals grey tones: While tech workers are more closely aligned with the new and cosmopolitan middle class, they also carry elements of the communitarian and ordinariness-seeking old middle class. The aim of this section is to explore the nuance against the backdrop of broad-stroked abstractions.
2. It's important to note that morality can be understood as a cultural principle in the sense that it constitutes a subjective meaning system. Building on Lamont (1992), however, I hold it is insightful to distinguish between cultural boundary-makings and moral boundary-makings as two different subjective meaning mechanisms that allocate resources and worth.
3. Also, it could be discussed whether the reorientations in the so-called great resignation that manifests in people quitting their jobs or reducing work hours indicates a return of social critique (Lichtenstein 2021).
4. Building on Sklair (2001), they can be considered part of the "technical fraction of global professionals" within the transnational capitalist class.
5. With regard to tech workers in Germany, this is not the case with EU citizens.
6. Tech hubs often emerge in places that offer liberal urban habitats with a close proximity to academic and financial institutions (Ferrary and Granovetter 2009; Saxenian 1996)
7. In certain countries of the Global South, however, the middle class has also come under pressure. This includes Egypt, where neoliberal reforms undermined material and political achievements of the middle class (Kandil 2012).

APPENDIX

1. The notion of "postindustrial" thus does not refer to a deindustrialized society but to a social space where industrial production has lost its once dominating economic force.
2. I relied on the rankings of the "Times Higher Education World University Ranking 2018" for the sampling. Certainly, rankings are controversial. However, they matter in the sense that they constitute an objectified reputation system that actors turn to for orientation.

References

Abbott, Andrew. 1988. *The System of Professions*. University of Chicago Press.

Abbott, Andrew. 2016. *Processual Sociology*. University of Chicago Press.

Acs, Zoltan J., and Ronnie J. Phillips. 2002. "Entrepreneurship and Philanthropy in American Capitalism." *Small Business Economics* 19 (3): 189–204.

AI Safety. 2023. "Statement on AI Risk | CAIS." https://www.safe.ai/statement-on-ai-risk#open-letter.

Akrich, Madeline. 1992. "The De-Inscription of Technical Objects." In *Shaping Technology/Building Society: Studies in Sociotechnical Change*, edited by Wiebe Bijker and John Law, 205–24. MIT Press.

Alexander, Jeffrey C. 2004. "Cultural Pragmatics: Social Performance Between Ritual and Strategy." *Sociological Theory* 22 (4): 527–73.

Alphabet Workers Union. 2023. *AWU Homepage*. Alphabet Workers Union. https://alphabetworkersunion.org/.

Altenried, Moritz, and Valentin Niebler. 2022. "Fragmentierte arbeit, verallgemeinerter konflikt: Alltägliche auseinandersetzungen in der plattformarbeit." In *Widerstand im arbeitsprozess eine arbeitssoziologische einführung*, edited by Heiner Heiland and Simon Schaupp, 277–300. Transcript.

Amrute, Sareeta. 2014. "Proprietary Freedoms in an IT Office: How Indian IT Workers Negotiate Code and Cultural Branding." *Social Anthropology/Anthropologie Sociale* 22 (1): 101–17.

Amrute, Sareeta. 2016. *Encoding Race, Encoding Class: Indian IT Workers in Berlin*. Duke University Press.

Amrute, Sareeta. 2019. "Of Techno-Ethics and Techno-Affects." *Feminist Review* 123 (1): 56–73.

Anderson, Elijah. 2000. *Code of the Street: Decency, Violence, and the Moral Life of the Inner City*. W. W. Norton & Company.

Andrew, Edward. 1983. "Class in Itself and Class Against Capital: Karl Marx and His Classifiers." *Canadian Journal of Political Science / Revue Canadienne de Science Politique* 16 (3): 577–84.

Atkinson, Anthony B., and Andrea Brandolini. 2013. "On the Identification of the Middle Class." In *Income Inequality: Economic Disparities and the Middle Class in Affluent Countries*, edited by Janet Gornick and Markus Jäntti, 77–100. Stanford University Press.

August, Vincent. 2021. "Technologisches regieren: Der aufstieg des netzwerk-denkens." In *Der krise der moderne. Foucault, Luhmann und die kybernetik.* Transcript.

Avnoon, Netta. 2021. "Data Scientists' Identity Work: Omnivorous Symbolic Boundaries in Skills Acquisition." *Work, Employment and Society* 35 (2): 332–49.

Baert, Patrick. 2022. "The Theory of Everything: A Sympathetic Critique of Andeas Reckwitz's *The Society of Singularities.*" *Analyse & Kritik* 44 (2): 323–29.

Badiou, Alain. 2007. *Being and Event.* Continuum.

Bainbridge, Wilma Alice, and William Sims Bainbridge. 2007. "Creative Uses of Software Errors: Glitches and Cheats." *Social Science Computer Review* 25 (1): 61–77.

Barbrook, Richard, and Andy Cameron. 1995. "The Californian Ideology." *Mute.* http://www.metamute.org/editorial/articles/californian-ideology.

Baron, Ethan. 2018. "H-1B: Foreign Citizens Make Up Nearly Three-Quarters of Silicon Valley Tech Workforce, Report Says." *The Mercury News.* https://www.mercurynews.com/2018/01/17/h-1b-foreign-citizens-make-up-nearly-three-quarters-of-silicon-valley-tech-workforce-report-says/.

Bechky, Beth A. 2003. "Object Lessons: Workplace Artifacts as Representations of Occupational Jurisdiction." *American Journal of Sociology* 109 (3): 720–52.

Beck, Linda, and Linus Westheuser. 2022. "Verletzte ansprüche. Zur grammatik des politischen bewusstseins von arbeiterinnen." *Berliner Journal für Soziologie* 32 (2): 279–316.

Beck, Ulrich. 1992. *Risk Society: Towards a New Modernity.* Sage.

Beecham, Sarah, Nathan Baddoo, Tracy Hall, Hugh Robinson, and Helen Sharp. 2008. "Motivation in Software Engineering: A Systematic Literature Review." *Information and Software Technology* 50 (9–10): 860–78.

Bell, Daniel. 1973. *The Coming of Post-Industrial Society: A Venture in Social Forecasting.* Basic Books.

Beltran, Hector. 2018. "Hacking Imaginaries: Codeworlds and Code Work Across the US/Mexico Borderlands." Diss., University of California, Berkeley.

Benjamin, Ruha. 2019. *Race After Technology: Abolitionist Tools for the New Jim Code.* Polity.

Bergvall-Kåreborn, Birgitta, and Debra Howcroft. 2013. "'The Future's Bright, the Future's Mobile': A Study of Apple and Google Mobile Application Developers." *Work, Employment and Society* 27 (6): 964–81.

Bertelsmann Stiftung. 2021. "Bröckelt die mittelschicht? Risiken und chancen für mittlere einkommensgruppen auf dem deutschen arbeitsmarkt." https://www.bertelsmann-stiftung.de/de/publikationen/publikation/did/broeckelt-die-mittelschicht-all.

Bhabha, Homi K. 1994. *The Location of Culture.* Routledge.

Blackwell, Alan. 2024. *Moral Codes: Designing Alternatives to AI.* MIT Press.

Boag, William, Harini Suresh, Bianca Lepe, and Catherine D'Ignazio. 2022. "Tech Worker Organizing for Power and Accountability." In *Proceedings of the 2022 ACM Conference on Fairness, Accountability, and Transparency*, 452–63. FAccT '22. Association for Computing Machinery. https://dl.acm.org/doi/10.1145/3531146.3533111.

Boes, Andreas, Tobias Kämpf, Barbara Langes, and Thomas Lühr. 2018. *Lean und agil im büro: Neue organisationskonzepte in der digitalen transformation und ihre folgen für die angestellten.* Transcript Verlag.

Boltanski, Luc. 2011. *On Critique: A Sociology of Emancipation.* Polity.

Boltanski, Luc, and Eve Chiapello. 2018. *The New Spirit of Capitalism.* Verso.

Bosančić, Sasa, Folke Brodersen, Lisa Pfahl, Lena Schürmann, Tina Spies, and Boris Traue. 2021. *Following the Subject. Studies in Subjectivation, Band 1: Grundlagen Und Forschungspraxis.* Springer.

Bourdieu, P. 2003. *Backfire: Against the Tyranny of the Market.* The New Press.

Bourdieu, Pierre. 1984. *Distinction: A Social Critique of the Judgement of Taste.* Harvard University Press.

Bourdieu, Pierre. 1985. "The Social Space and the Genesis of Groups." *Theory and Society* 14 (6): 723–44.

Bourdieu, Pierre. 1987. "What Makes a Social Class? On The Theoretical and Practical Existence of Groups." *Berkeley Journal of Sociology* 32:1–17.

Bourdieu, Pierre. 1988. *Homo Academicus.* Stanford University Press.

Bourdieu, Pierre. 1990. *The Logic of Practice.* Stanford University Press.

Bourdieu, Pierre. 1996. *The Rules of Art: Genesis and Structure of the Literary Field.* Stanford University Press.

Bourdieu, Pierre. 1998. *Practical Reason: On the Theory of Action.* Stanford University Press.

Bourdieu, Pierre. 2005. "Principles of an Economic Anthropology." In *The Handbook of Economic Sociology*, edited by Neil J. Smelser and Richard Swedberg, 75–89. Princeton University Press.

Bourdieu, Pierre. 2008. *Sketch for a Self-Analysis.* Polity Press.

Bourdieu, Pierre, and Loïc J. D. Wacquant. 1992. *An Invitation to Reflexive Sociology.* University of Chicago Press.

Boxell, Levi, Matthew Gentzkow, and Jesse M. Shapiro. 2024. "Cross-Country Trends in Affective Polarization." *The Review of Economics and Statistics* 106 (2): 557–65.

Braverman, Harry. 1974. *Labor and Monopoly Capital: The Degradation of Work in the Twentieth Century.* Monthly Review Press.

Brenner, Johanna, and Nancy Fraser. 2017. "What Is Progressive Neoliberalism?: A Debate." *Dissent* 64 (2): 130–130.

Brock, Andre. 2015. "Deeper Data: A Response to Boyd and Crawford." *Media, Culture & Society* 37 (7): 1084–88.

Bröckling, Ulrich. 2015. *The Entrepreneurial Self: Fabricating a New Type of Subject.* Sage.

Brorsen Smidt, Thomas, Fredrik Bondestam, Gyða Margrét Pétursdóttir, and Þorgerður Einarsdóttir. 2020. "Expanding Gendered Sites of Resistance in the Neoliberal Academy." *European Journal of Higher Education* 10 (2): 115–29.

Browne, Jude, Eleanor Drage, and Kerry McInerney. 2024. "Tech Workers' Perspectives on Ethical Issues in AI Development: Foregrounding Feminist Approaches." *Big Data & Society* 11 (1): 1-11.

Brown, Wendy. 2015. *Undoing the Demos. Neoliberalism's Stealth Revolution.* Princeton University Press.

Burgess, Jean. 2012. *The iPhone Moment, the Apple Brand and the Creative Consumer: From "Hackability and Usability" to Cultural Generativity.* Routledge.

Burrell, Jenna, and Marion Fourcade. 2021. "The Society of Algorithms." *Annual Review of Sociology* 47: 213–37.

Butler, Judith. 2005. *Giving an Account of Oneself.* Fordham University Press.

Callon, Michel. 1984. "Some Elements of a Sociology of Translation: Domestication of the Scallops and the Fishermen of St Brieuc Bay." *Sociological Review* 32 (1): 196–233.

Callon, Michel. 1998. "Introduction: The Embeddedness of Economic Markets in Economics." *Sociological Review* 46 (1): 1–57.

Callon, Michel. 2024. "Working with Bruno Latour on a Daily Basis." *Theory, Culture & Society* 41 (5): 1–19.

Cambon, Sarah Chaney, and Gwynn Guilford. 2022. "WSJ News Exclusive | Laid Off Tech Workers Quickly Find New Jobs." *Wall Street Journal,* December 27, sec. Economy. https://www.wsj.com/articles/laid-off-tech-workers-quickly-find-new-jobs-11672097730.

Carrigan, Coleen. 2024. *Cracking the Bro Code.* MIT Press.

Chen, Carolyn. 2022. *Work Pray Code: When Work Becomes Religion in Silicon Valley.* Princeton University Press.

Chiapello, Eve. 2013. "Capitalism and Its Criticisms." In *New Spirits of Capitalism? Crises, Justifications, and Dynamics,* edited by Paul Du Gay and Glenn Morgan, 60–81. Oxford University Press.

Childress, Clayton, Shyon Baumann, Craig M. Rawlings, and Jean-François Nault. 2021. "Genres, Objects, and the Contemporary Expression of Higher-Status Tastes." *Sociological Science* 8: 230–64.

Chouliaraki, Lilie. 2013. *The Ironic Spectator: Solidarity in the Age of Post-Humanitarianism.* Polity.

Christiaens, Tim. 2020. "Digital Biopolitics and the Problem of Fatigue in Platform Capitalism." In *Big Data—A New Medium?,* edited by Natasha Lushetich, 80–93. Routledge.

Christiaens, Tim. 2022. "Convivial Autonomy in Platform Capitalism." In *New Interdisciplinary Perspectives on and Beyond Autonomy,* edited by Christopher Watkin and Oliver Davis, 69–82. Taylor & Francis.

Christin, Angèle. 2020. *Metrics at Work: Journalism and the Contested Meaning of Algorithms.* Princeton University Press.

Chua, Phoebe K., and Melissa Mazmanian. 2020. "Are You One of Us? Current Hiring Practices Suggest the Potential for Class Biases in Large Tech Companies." *Proceedings of the ACM on Human-Computer Interaction* 4 (CSCW2). ACM: 1–20.

Chun, Wendy Hui Kyong. 2021. *Discriminating Data: Correlation, Neighborhoods, and the New Politics of Recognition.* MIT Press.

CompTIA. 2025. *State of the Tech Workforce.* The Computer Technology Industry Association (CompTIA). https://www.comptia.org/en/resources/research/state-of-the-tech-workforce-2025/

Couldry, Nick, and Ulises A. Mejias. 2019. *The Costs of Connection: How Data Is Colonizing Human Life and Appropriating It for Capitalism.* Stanford University Press.

Creech, Brian, and Perry Parks. 2022. "Promises Granted: Venture Philanthropy and Tech Ideology in Metajournalistic Discourse." *Journalism Studies* 23 (1): 70–88.

Davenport, Thomas H., and D. J. Patil. 2012. "Data Scientist." *Harvard Business Review* 90 (5): 70–76.

Dawson, Ashley. 2002. "'This Is the Digital Underclass': Asian Dub Foundation and Hip-Hop Cosmopolitanism." *Social Semiotics* 12, (1): 27–44.

Dencik, Lina, Emiliano Treré, Joanna Redden, and Arne Hintz. *Data Justice.* Sage.

Deleuze, Gilles. 1994. *Difference and Repetition.* Columbia University Press.

Deleuze, Gilles, and Félix Guattari. 1987. *A Thousand Plateaus: Capitalism and Schizophrenia.* University of Minnesota Press.

Di, Di. 2023. "Ethical Ambiguity and Complexity: Tech Workers' Perceptions of Big Data Ethics in China and the US." *Information, Communication & Society* 26 (5): 957–73.

Dörre, Klaus, Sophie Bose, John Lütten, and Jakob Köster. 2018. "Arbeiterbewegung von rechts? Motive und grenzen einer imaginären revolte." *Berliner Journal Für Soziologie* 28 (1): 55–89.

Dorschel, Robert. 2021. "Discovering Needs for Digital Capitalism: The Hybrid Profession of Data Science." *Big Data & Society* 8 (2): 1–13.

Dorschel, Robert. 2022a. "A New Middle-Class Fraction with a Distinct Subjectivity: Tech Workers and the Transformation of the Entrepreneurial Self." *The Sociological Review* 70 (6): 1302–20.

Dorschel, Robert. 2022b. "Reconsidering Digital Labour: Bringing Tech Workers into the Debate." *New Technology, Work and Employment* 37 (2): 288–307.

Dorschel, Robert. 2024. "Middle-Class Responses to Climate Change: An Analysis of the Ecological Habitus of Tech Workers." *Current Sociology* 72 (7): 1301–18.

Doyuran, Elif Buse. 2023. "Nudge Goes to Silicon Valley: Designing for the Disengaged and the Irrational." *Journal of Cultural Economy* 17 (1): 1–19.

Dunbar-Hester, Christina. 2019. *Hacking Diversity: The Politics of Inclusion in Open Technology Cultures*. Princeton University Press.

Durkheim, Emile. 1997. *The Division of Labor in Society*. Free Press.

Dyer-Witheford, Nick. 2015. *Cyber-Proletariat: Global Labour in the Digital Vortex*. Pluto Press.

Ehrenberg, Alain. 2010. *Weariness of the Self: Diagnosing the History of Depression in the Contemporary Age*. McGill-Queen University Press.

Ehrenreich, Barbara, and John Ehrenreich. 1979. "The Professional-Managerial Class." In *Between Labor and Capital*, edited by Pat Walker, 5:5–45. South End Press.

Esping-Andersen, Gosta. 1990. *The Three Worlds of Welfare Capitalism*. Princeton University Press.

Facebook. 2022. "Data for Good: Startseite." https://dataforgood.facebook.com/.

Fanon, Frantz. 2008. *Black Skin, White Masks*. Grove Press.

Ferrary, Michel, and Mark Granovetter. 2009. "The Role of Venture Capital Firms in Silicon Valley's Complex Innovation Network." *Economy and Society* 38 (2): 326–59.

Florida, Richard. 2002. *The Rise of the Creative Class and How It's Transforming Work, Leisure, Community and Everyday Life*. Basic Books.

Florida, Richard. 2010. *Who's Your City? How the Creative Economy Is Making Where to Live the Most Important Decision of Your Life*. Vintage Canada.

Foucault, Michel. 1978. *Dispositive der macht: Über sexualität, wissen und wahrheit*. Merve-Verlag.

Foucault, Michel. 1982. "The Subject and Power." *Critical Inquiry* 8 (4): 777–95.

Foucault, Michel. 1988. *Technologies of the Self: A Seminar with Michel Foucault*. University of Massachusetts Press.

Foucault, Michel. 1995. *Discipline and Punish: The Birth of the Prison*. Vintage Books.

Foucault, Michel. 2002. *The Archaeology of Knowledge*. Routledge.

Foucault, Michel. 2007. *Security, Territory, Population: Lectures at the Collège de France*. 1977–78. Palgrave Macmillan.

Foucault, Michel. 2008. *The Birth of Biopolitics—Lectures at the Collège de France, 1978–1979*. Palgrave Macmillan.

Foucault, Michel. 2012a. *The History of Sexuality 2: The Use of Pleasure*. Vintage Books.

Foucault, Michel. 2012b. *The History of Sexuality, Volume 3: The Care of the Self*. Vintage Books.

Fourcade, Marion, and Kieran Healy. 2013. "Classification Situations: Life-Chances in the Neoliberal Era." *Accounting, Organizations and Society* 38 (8): 559–72.

Fourcade, Marion, and Kieran Healy. 2024. *The Ordinal Society*. Harvard University Press.

Franke, Marcus. 2023. "Interview with Gianpaolo Meloni from the Amazon EWC: Five Questions for the EWC Secretary on Transnational Solidarity in Working Relations." *Journal of Political Sociology* 1 (2): 160–64.

Fraser, Nancy. 2009. "Social Justice in the Age of Identity Politics." In *Geographic Thought: A Praxis Perspective*, edited by George Henderson and Marvin Waterstone, 72–91. Routledge.

Freidson, Eliot. 1988. *Profession of Medicine*. University of Chicago Press.

Frey, Carl Benedikt, and Michael A. Osborne. 2017. "The Future of Employment: How Susceptible Are Jobs to Computerisation?" *Technological Forecasting and Social Change* 114: 254–80.

Friedman, Sam. 2012. "Cultural Omnivores or Culturally Homeless? Exploring the Shifting Cultural Identities of the Upwardly Mobile." *Poetics* 40 (5): 467–89.

Friedman, Sam. 2016. "Habitus Clivé and the Emotional Imprint of Social Mobility." *The Sociological Review* 64 (1): 129–47.

Friedman, Sam, Christoph Ellersgaard, Aaron Reeves, and Anton Grau Larsen. 2024. "The Meaning of Merit: Talent Versus Hard Work Legitimacy." *Social Forces* 102 (3): 861–79.

Friedman, Sam, and Aaron Reeves. 2020. "From Aristocratic to Ordinary: Shifting Modes of Elite Distinction." *American Sociological Review* 85 (2): 323–50.

Gabriel, Iason. 2017. "Effective Altruism and Its Critics." *Journal of Applied Philosophy* 34 (4): 457–73.

Gabrys, Jennifer. 2011. *Digital Rubbish: A Natural History of Electronics*. University of Michigan Press.

Geiger, S. 2019. "Theorizing Disruption Through an Eschatological Lens: Silicon Valley, Inevitability and the End of History." *Journal of Cultural Economy* 13 (2): 169–84.

Gerbaudo, Paolo. 2018. *The Digital Party*. Pluto Press.

Giddens, Anthony. 1991. *Modernity and Self-Identity: Self and Society in the Late Modern Age*. Stanford University Press.

Glaser, Barney, and Anselm Strauss. 1967. *The Discovery of Grounded Theory: Strategies for Qualitative Research*. Aldine de Gruyter.

Goffman, Erving. 1961. *Asylums: Essays on the Social Situation of Mental Patients and Other Inmates*. Anchor Books.

Goldstein, Jan. 1984. "Foucault Among the Sociologists: The" Disciplines" and the History of the Professions." *History and Theory* 23 (2): 170–92.

Goldthorpe, John H., David Lockwood, Frank Bechhofer, and Jennifer Platt. 1967. "The Affluent Worker and the Thesis of Embourgeoisement: Some Preliminary Research Findings." *Sociology* 1 (1): 11–31.

Graham, Mark, Isis Hjorth, and Vili Lehdonvirta. 2017. "Digital Labour and Development: Impacts of Global Digital Labour Platforms and the Gig Economy on Worker Livelihoods." *Transfer: European Review of Labour and Research* 23 (2): 135–62.

Gramsci, Antonio. 1971. *Selections from the Prison Notebooks*. Lawrence & Wishart.

Grasmeier, Marie C., and Linda Beck. 2023. "Boundary-Work, Occupational Identities and Class-Experiences of Global Seafarers." In *Maritime Professions: Issues and Perspectives*, edited by Agnieszka Kołodziej-Durnaś, Frank Sowa, and Marie C. Grasmeier, 2:58–92. Brill.

Gray, Mary L., and Siddharth Suri. 2019. *Ghost Work: How to Stop Silicon Valley from Building a New Global Underclass*. Houghton Mifflin Harcourt.

Gregg, Melissa. 2012. "The Return of Organisation Man: Commuter Narratives and Suburban Critique." *Cultural Studies Review* 18 (2): 242–61.

Grohmann, Rafael. 2023. "Not Just Platform, nor Cooperatives: Worker-Owned Technologies from Below." *Communication, Culture and Critique* 16 (4): 274–82.

Gupta, Hemangini. 2019. "Testing the Future: Gender and Technocapitalism in Start-up India." *Feminist Review* 123 (1): 74–88.

Hall, Peter A., and Michèle Lamont. 2013. *Social Resilience in the Neoliberal Era*. Cambridge University Press.

Hall, Peter A., and David Soskice. 2001. *Varieties of Capitalism. The Institutional Foundations of Comparative Advantage*. Oxford University Press.

Hall, Stuart. 2019. *Essential Essays, Volume 1: Foundations of Cultural Studies*. Edited by David Morley. Stuart Hall: Selected Writings. Duke University Press.

Hall, Stuart, and Paul Du Gay. 1996. *Questions of Cultural Identity*. Sage.

Hanlon, Gerard, and Prof Peter Fleming. 2009. "Updating the Critical Perspective on Corporate Social Responsibility." *Sociology Compass* 3 (6): 937–48.

Hardt, Michael, and Antonio Negri. 2000. *Empire*. Harvard University Press.

Hicks, Mar. 2017. *Programmed Inequality: How Britain Discarded Women Technologists and Lost Its Edge in Computing*. MIT Press.

Hochschild, Arlie. 1983. *The Managed Heart: Commercialization of Human Feeling*. University of California Press.

Hochschild, Arlie Russell. 1979. "Emotion Work, Feeling Rules, and Social Structure." *American Journal of Sociology* 85 (3): 551–75.

Hochschild, Arlie Russell. 2018. *Strangers in Their Own Land: Anger and Mourning on the American Right*. The New Press.

Hofstätter, Lukas, Marco Hohmann, Sighard Neckel, and Conny Petzold. 2016. "Researching the Global Financial Class. Working Paper." http://publikationen.ub.uni-frankfurt.de/opus4/frontdoor/index/index/docId/44443.

Huber, Matthew T. 2022. *Climate Change as Class War: Building Socialism on a Warming Planet*. Verso Books.

Huws, Ursula. 2001. "The Making of a Cybertariat? Virtual Work in a Real World." *Socialist Register* 37:1–23.

Illich, Ivan. 1973. *Tools for Conviviality*. Harper & Row.

Illouz, Eva. 2007. *Cold Intimacies: The Making of Emotional Capitalism*. Polity.

Ingram, Nicola, and Kim Allen. 2019. "'Talent-Spotting' or 'Social Magic'? Inequality, Cultural Sorting and Constructions of the Ideal Graduate in Elite Professions." *The Sociological Review* 67 (3): 723–40.

IPCC. 2022. *IPCC Sixth Assessment Report: Impacts, Adaptation and Vulnerability*. https://www.ipcc.ch/report/ar6/wg2/.

Irani, Lilly. 2015a. "Difference and Dependence Among Digital Workers: The Case of Amazon Mechanical Turk." *South Atlantic Quarterly* 114 (1): 225–34.

Irani, Lilly. 2015b. "Hackathons and the Making of Entrepreneurial Citizenship." *Science, Technology, & Human Values* 40 (5): 799–824.

Irani, Lilly. 2019. *Chasing Innovation: Making Entrepreneurial Citizens in Modern India*. Princeton University Press.

Jarness, Vegard, and Magne Flemmen. 2019. "A Struggle on Two Fronts: Boundary Drawing in the Lower Region of the Social Space and the Symbolic Market for 'Down-to-Earthness.'" *British Journal of Sociology* 70 (1), 166–189.

Jarrett, Kylie. 2022. *Digital Labor*. Polity Press.

Kalberg, Stephen. 2004. "The Past and Present Influence of World Views: Max Weber on a Neglected Sociological Concept." *Journal of Classical Sociology* 4 (2): 139–63.

Kandil, Hazem. 2012. "Why Did the Egyptian Middle Class March to Tahrir Square?" *Mediterranean Politics* 17 (2): 197–215.

Kapoor, Sayash, Matthew Sun, Mona Wang, Klaudia Jazwinska, and Elizabeth Anne Watkins. 2022. "Weaving Privacy and Power: On the Privacy Practices of Labor Organizers in the U.S. Technology Industry." *Proceedings of the ACM on Human-Computer Interaction* 6 (CSCW2): 473:1–473:33.

Karppi, Tero, Urs Stäheli, Clara Wieghorst, and Lea P. Zierott. 2021. *Undoing Networks*. Meson Press.

Keller, Reiner. 2004. "Der Müll Der Gesellschaft. Eine Wissenssoziologische Diskursanalyse." In *Handbuch Sozialwissenschaftliche Diskursanalyse*, edited by R. Keller, A. Hierseland, W. Schneider, and W. Viehöver, 197–232. Springer.

Keller, Reiner. 2011. "The Sociology of Knowledge Approach to Discourse (SKAD)." *Human Studies* 34 (1): 43–65.

Keller, Reiner. 2013. *Doing Discourse Research: An Introduction for Social Scientists*. Sage.

Kellogg, Katherine C., Melissa A. Valentine, and Angèle Christin. 2019. "Algorithms at Work: The New Contested Terrain of Control." *Academy of Management Annals* 14 (1): 366–410.

Kendall, Lori. 1999. "Nerd Nation: Images of Nerds in US Popular Culture." *International Journal of Cultural Studies* 2 (2): 260–83.

Khan, Shamus, and Colin Jerolmack. 2013. "Saying Meritocracy and Doing Privilege." *The Sociological Quarterly* 54 (1): 9–19.

Klammer, Ute. 2023. "The Ambivalent Trajectory of the German Gender Regime: Are Female Breadwinner Families an Indicator of a Shift Towards a Public Gender Regime?" *Women's Studies International Forum* 99 (3): 1–8.

Koopmans, Ruud, and Michael Zürn. 2019. "Cosmopolitanism and Communitarianism—How Globalization Is Reshaping Politics in the Twenty-First Century." In *The Struggle over Borders: Cosmopolitanism and Communitarianism*, edited by Pieter De Wilde, Ruud Koopmans, Wolfgang Merkel, and Michael Zürn, 1–34. Cambridge University Press.

Kumkar, Nils, and Uwe Schimank. 2021. "Drei-klassen-gesellschaft? Bruch? Konfrontation? Eine auseinandersetzung mit Andreas Reckwitz'diagnose der spätmoderne." *Leviathan* 49 (1): 7–32.

Kunda, Gideon. 1992. *Engineering Culture: Control and Commitment in a High-Tech Corporation.* Temple University Press.

Lahire, Bernard. 2011. *The Plural Actor.* Polity.

Lamont, Michèle. 1992. *Money, Morals, and Manners: The Culture of the French and the American Upper-Middle Class.* University of Chicago Press.

Lamont, Michèle. 2002. *The Dignity of Working Men.* Harvard University Press.

Lamont, Michèle. 2023. *Seeing Others: How Recognition Works—and How It Can Heal a Divided World.* Simon and Schuster.

Lamont, Michèle, and Sada Aksartova. 2002. "Ordinary Cosmopolitanisms: Strategies for Bridging Racial Boundaries Among Working Class Men." *Theory, Culture and Society* 19 (4): 1–25.

Lamont, Michèle, and Virág Molnár. 2002. "The Study of Boundaries in the Social Sciences." *Annual Review of Sociology* 28 (1): 167–95.

Lamont, Michèle, and Ann Swidler. 2014. "Methodological Pluralism and the Possibilities and Limits of Interviewing." *Qualitative Sociology* 37 (2): 153–71.

Latour, Bruno. 2005. *Reassembling the Social: An Introduction to Actor-Network-Theory.* Oxford University Press.

Leggett, William. 2021. "Can Mindfulness Really Change the World? The Political Character of Meditative Practices." *Critical Policy Studies* 16 (3): 261–78.

Lehdonvirta, Vili. 2018. "Flexibility in the Gig Economy: Managing Time on Three Online Piecework Platforms." *New Technology, Work and Employment* 33 (1): 13–29.

Lehuedé, Sebastián. 2025. "An Elemental Ethics for Artificial Intelligence: Water as Resistance Within AI's Value Chain." *AI & SOCIETY* 40:1761–1774.

Lichtenstein, Nelson. 2021. "Perspective | Are We Witnessing a "General Strike" in Our Own Time?" *Washington Post,* November 18. https://www.washingtonpost.com/outlook/2021/11/18/are-we-witnessing-general-strike-our-own-time/.

Liu, Hong Yu. 2022. "Digital Taylorism in China's e-Commerce Industry: A Case Study of Internet Professionals." *Economic and Industrial Democracy* 44 (1): 262–79.

Lobo, Rita. 2014. "Silicon Valley's Sexist Brogrammer Culture Is Locking Women Out of Tech." *The New Economy*. https://www.theneweconomy.com/technology/silicons-sexist-brogrammer-culture-is-locking-women-out-of-tech.

Lupu, Ioana, and Laura Empson. 2015. "Illusio and Overwork: Playing the Game in the Accounting Field." Edited by Professor Crawford Spence and Professor Daniel Muzio Professor Chris Carter. *Accounting, Auditing & Accountability Journal* 28 (8): 1310–40.

MacKenzie, Donald, and Judy Wajcman. 1999. *The Social Shaping of Technology*. Open University Press.

Marcuse, Herbert. 2013. *One-Dimensional Man: Studies in the Ideology of Advanced Industrial Society*. Routledge.

Marwick, Alice E. 2013. *Status Update: Celebrity, Publicity and Branding in the Social Media Age*. Yale University Press.

Marx, Karl. 1963. *Eighteenth Brumaire of Louis Bonaparte*. International Publishers.

Mason, Paul. 2016. *Postcapitalism: A Guide to Our Future*. Macmillan.

Mau, Steffen. 2015. *Inequality, Marketization and the Majority Class: Why Did the European Middle Classes Accept Neo-Liberalism?* Palgrave Pivot.

Mau, Steffen, Thomas Lux, and Fabian Gülzau. 2020. "Die drei Arenen der neuen Ungleichheitskonflikte. Eine sozialstrukturelle Positionsbestimmung der Einstellungen zu Umverteilung, Migration und sexueller Diversität." *Berliner Journal für Soziologie* 30 (3): 317–46.

Maxwell, Joseph. 2012. *Qualitative Research Design. An Interactive Approach*. Sage.

McPherson, Ella. 2023. "Witnessing: Iteration and Social Change." *AI & SOCIETY* 38 (5): 1987–95.

Meghji, Ali. 2019. *Black Middle Class Britannia: Identities, Repertoires, Cultural Consumption*. Manchester University Press.

Meghji, Ali, and Rima Saini. 2018. "Rationalising Racial Inequality: Ideology, Hegemony and Post-Racialism Among the Black and South Asian Middle-Classes." *Sociology* 52 (4): 671–87.

Merton, Robert K. 1936. "The Unanticipated Consequences of Purposive Social Action." *American Sociological Review* 1 (6): 894–904.

Mezzadra, Sandro, and Brett Neilson. 2013. *Border as Method, or, the Multiplication of Labor*. Duke University Press.

Milanovic, Branko. 2016. *Global Inequality*. Harvard University Press.

Molinari, Carmen. 2020. "There Is Something Missing from Tech Worker Organizing." *Organizing.Work*. December 9. https://organizing.work/2020/12/there-is-something-missing-from-tech-worker-organizing/.

Morozov, Evgeny. 2013. *To Save Everything, Click Here: The Folly of Technological Solutionism*. Public Affairs.

Müller, Annekathrin. 2023. "Community als ware: Eine ethnographische untersuchung der "Factory Berlin" als urbanem arbeitsort der innovationsproduktion." *Sub\urban. zeitschrift für kritische stadtforschung* 11 (1/2): 127–48.

Müller, Hans-Peter, and Steffen Sigmund. 2014. *Max Weber Handbuch: Leben—Werk—Wirkung.* J. B. Metzler.

Muzio, Daniel, Damian Hodgson, James Faulconbridge, Jonathan Beaverstock, and Sarah Hall. 2011. "Towards Corporate Professionalization: The Case of Project Management, Management Consultancy and Executive Search." *Current Sociology* 59 (4): 443–64.

Nachtwey, Oliver. 2016. *Die Abstiegsgesellschaft. Über Das Aufbegehren in Der Regressiven Moderne.* Suhrkamp.

Nachtwey, Oliver, and Simon Schaupp. 2022. "Ungleicher gabentausch—User-interaktionen und wertschöpfung auf digitalen plattformen." *Kölner Zeitschrift für Soziologie und Sozialpsychologie* 74: 59–80.

Nachtwey, Oliver, and Timo Seidl. 2020. "The Solutionist Ethic and the Spirit of Digital Capitalism." University of Basel; SocArXiv. Accessed July 10, 2020. https://edoc.unibas.ch/76426/1/Nachtwey%2C%20Seidl_The%20Solutionist%20Ethic%20and%20the%20Spirit%20of%20Digital%20Capitalism.pdf.

Neckel, Sighard. 2017. "The Sustainability Society: A Sociological Perspective." *Culture, Practice & Europeanization* 2 (2): 46–52.

Neckel, Sighard, Lukas Hofstätter, and Marco Hohmann. 2018. *Die Globale Finanzklasse: Business, Karriere, Kultur in Frankfurt Und Sydney*. Campus Verlag.

Neckel, Sighard, and Greta Wagner. 2013. *Leistung Und Erschöpfung: Burnout in Der Wettbewerbsgesellschaft.* Suhrkamp Verlag.

Nedzhvetskaya, Nataliya, and J. S. Tan. 2024a. "Labor's Stake in Shaping Tech Futures." *Association for Computing Machinery* 31 (4): 50–53.

Nedzhvetskaya, Nataliya, and J. S. Tan. 2024b. "The Role of Workers in AI Ethics and Governance." In *The Oxford Handbook of AI Governance*, edited by Justin Bullock, Yu-Che Chen, Johannes Himmelreich, Valerie M. Hudson, Anton Korinek, Matthew M. Young, Baobao Zhang. Oxford University Press.

Neely, Megan Tobias. 2020. "The Portfolio Ideal Worker: Insecurity and Inequality in the New Economy." *Qualitative Sociology* 43 (2): 271–96.

Neely, Megan Tobias. 2022. *Hedged Out: Inequality and Insecurity on Wall Street.* University of California Press.

Neely, Megan Tobias, Patrick Sheehan, and Christine L. Williams. 2023. "Social Inequality in High Tech: How Gender, Race, and Ethnicity Structure the World's Most Powerful Industry." *Annual Review of Sociology* 49:319–38.

Neff, Gina. 2012. *Venture Labor: Work and the Burden of Risk in Innovative Industries.* MIT Press.

Neff, Gina, and Dawn Nafus. 2016. *Self-Tracking*. MIT Press.

Niebler, Valentin. 2020. "'YouTubers Unite': Collective Action by YouTube Content Creators." *Transfer: European Review of Labour and Research* 26 (2): 223–27.

Niebler, Valentin. 2023a. "Coalitional Power in the Digital Economy: Alliances of Gig and Tech Workers." In *Trajectories of Platform Capitalism and Platform Work*, 9–15. Friedrich-Ebert-Stiftung.

Niebler, Valentin. 2023b. "Transcending Borders? Horizons and Challenges of Global Tech Worker Solidarity." *Journal of Political Sociology* 1 (2): 57–76.

Niebler, Valentin. 2025. "What Is a Tech Worker, Really?: On the Meanings and Meaning-Making of a Term." *Work Organisation, Labour & Globalisation* 19:14–30.

Noble, David. 1984. *Forces of Production: A Social History of Industrial Automation*. Knopf.

Park, Robert E. 1928. "Human Migration and the Marginal Man." *American Journal of Sociology* 33 (6): 881–93.

Parsons, Talcott. 1939. "The Professions and Social Structure." *Social Forces* 17 (4): 457–67.

Patrick-Thomson, Holly, and Michael Kranert. 2021. "Don't Work for Free: Online Discursive Resistance to Precarity in Commercial Photography." *Work, Employment and Society* 35 (6): 1034–52.

Peterson, Richard A., and Roger M. Kern. 1996. "Changing Highbrow Taste: From Snob to Omnivore." *American Sociological Review* 61 (5): 900–907.

Pfadenhauer, Michaela. 2003. *Professionalität. Eine wissenssoziologische rekonstruktion institutionalisierter kompetenzdarstellungskompetenz*. Leske & Budrich.

Pfahl, Lisa, Lena Schürmann, and Boris Traue. 2018. "Subjektivierungsanalyse." In *Handbuch Interpretativ Forschen*, edited by Leila Akremi, Nina Baur, Hubert Koblauch, and Boris Traue, Beltz Verlag, 858–85. Weinheim.

Piketty, Thomas. 2014. *Capital in the Twenty-First Century*. Harvard University Press.

Piketty, Thomas. 2021. *Capital and Ideology*. Harvard University Press.

Pistor, Katharina. 2019. *The Code of Capital*. Princeton University Press.

Pollack, Detlef, and Gergely Rosta. 2017. *Religion and Modernity: An International Comparison*. Oxford University Press.

Pongratz, Hans J., and Guenther Voss. 2003. "From Employee to 'Entreployee': Towards a 'Self-Entrepreneurial' Work Force?" *Concepts and Transformation* 8 (3): 239–54.

Poster, Winifred R. 2013. "Subversions of Techno-Masculinity: Indian ICT Professionals in the Global Economy." In *Rethinking Transnational Men*, edited by Jeff Hearn, Marina Blagojević, and Katherine Harrison, 129–49. Routledge.

Poulantzas, Nicos. 1979. "The New Petty Bourgeoisie." *Insurgent Sociologist* 9 (1): 56–60.

Pressman, Steven. 2015. "Defining and Measuring the Middle Class." *American Institute for Economic Research Working Paper*, 1–27. https://aier.org/wp-content/uploads/2016/10/WP007-Middle-Class.pdf.

Prior, Nick. 2008. "Putting a Glitch in the Field: Bourdieu, Actor Network Theory and Contemporary Music." *Cultural Sociology* 2 (3): 301–19.

Purser, Ronald. 2019. *McMindfulness: How Mindfulness Became the New Capitalist Spirituality*. Repeater.

Putnam, Robert D. 2016. *Our Kids: The American Dream in Crisis*. Simon and Schuster.

Radhakrishnan, Smithaa. 2011. *Appropriately Indian: Gender and Culture in a New Transnational Class*. Duke University Press.

Ramtin, Ramin. 1991. *Capitalism and Automation: Revolution in Technology and Capitalist Breakdown*. Pluto Press.

Raza, Sebastian. 2022. "Max Weber and Charles Taylor: On Normative Aspects of a Theory of Human Action." *Journal of Classical Sociology* 23 (1): 1–40.

Reckwitz, Andreas. 2010. *Das Hybride Subjekt*. Velbrück.

Reckwitz, Andreas. 2020. *Society of Singularities*. John Wiley & Sons.

Reckwitz, Andreas. 2021. *The End of Illusions: Politics, Economy, and Culture in Late Modernity*. Polity Press.

Rider, Karina Alexis. 2021. "Volunteering the Valley: Designing Technology for the Common Good in the San Francisco Bay Area." PhD Diss., Queen's University.

Riesman, David, Nathan Glazer, and Reuel Denney. 2001. *The Lonely Crowd: A Study of the Changing American Character*. Yale University Press.

Ritzer, George. 1971. "Occupational Myths." *The Kansas Journal of Sociology* 7 (1): 5–16.

Rone, Julia. 2021. "Hacktivism." In *The Routledge Encyclopedia of Citizen Media*, edited by Luis Perez-Gonzalez, Mona Baker, Molette Blagaard, and Henry Jones, 190–96. Routledge.

Rosa, Hartmut. (2005) 2013. *Social Acceleration: A New Theory of Modernity*. Columbia University Press.

Rosa, Hartmut. 2019. *Resonance: A Sociology of Our Relationship to the World*. John Wiley & Sons.

Rose, Nikolas. 1989. *Governing the Soul: The Shaping of the Private Self*. Routledge.

Rosen, Jay. 2012. "The People Formerly Known as the Audience." In *The Social Media Reader*, edited by Michael Mandiberg, 13–17. New York University Press.

Ross, Andrew. 2003. *No-Collar: The Humane Workplace and Its Hidden Costs*. Temple University Press.

Rothstein, Sydney. 2022. *Recoding Power: Tactics for Mobilizing Tech Workers*. Oxford University Press.

Sainato, Michael. 2022. "New Wave of Teachers' Strikes Rolls Across US." *The Guardian*, sec. US news. https://www.theguardian.com/us-news/2022/mar/25/teachers-strikes-us-low-pay-covid.

Savage, Mike. 2015. *Social Class in the 21st Century*. Penguin UK.

Savage, Mike, Gaynor Bagnall, and Brian Longhurst. "Ordinary, Ambivalent and Defensive: Class Identities in the Northwest of England." *Sociology* 35 (4): 875–92.

Saxenian, AnnaLee. 1996. *Regional Advantage: Culture and Competition in Silicon Valley and Route 128*. Harvard University Press.

Sayer, Andrew. 2005. *The Moral Significance of Class*. Cambridge University Press.

Schelsky, Helmut. 1967. *Wandlungen Der Deutschen Familie in Der Gegenwart*. Enke Verlag.

Schmidt-Wellenburg, Christian. 2013. *Die Regierung Des Unternehmens: Managementberatung Im Neoliberalen Kapitalismus* [The governmentality of the enterprise: Management consulting in neoliberal capitalism]. Herbert von Halem Verlag.

Schmidt-Wellenburg, Christian. 2014. "Der aufstieg der beratung zur transnationalen regierungsform im feld des managements." *Berliner Journal Für Soziologie* 24 (2): 227–55.

Scholz, Trebor. 2016. *Platform Cooperativism: Challenging the Corporate Sharing Economy*, 1–32. Rosa Luxemburg Foundation.

Sedgwick, Eve Kosofsky. 2003. "*Touching Feeling*. Affect, Pedagogy, Performativity." Duke University Press.

Selling, Niels, and Pontus Strimling. 2023. "Liberal and Anti-Establishment: An Exploration of the Political Ideologies of American Tech Workers." *The Sociological Review* 71 (6): 1467–97.

Sengupta, Subhasree, and Zhasmina Tacheva. 2022. "'Digital Sanctums of Empowerment': Exploring Community and Everyday Resilience-Building Tactics in Online Professional Communities for Women." *Companion Publication of the 2022 Conference on Computer Supported Cooperative Work and Social Computing*, 89–93.

Sennett, Richard. 2007. *The Culture of the New Capitalism*. Yale University Press.

Seto, Kenzo Soares. 2025. "Voices of Brazilian Tech Workers : Emerging Collective Actions in Brazil's Tech Community." *Work Organisation, Labour & Globalisation* 19:50–66.

Shakthi, S. 2023a. "Corporate Brahminism and Tech Work: Caste in a Modern Indian Profession." *South Asia: Journal of South Asian Studies* 46 (5): 920–33.

Shakthi, S. 2023b. "Travelling 'Down South': Language, Cultural Capital and Spatiality in Chennai's Information Technology Sector." *Social & Cultural Geography* 25 (3): 478–95.

Sheehan, Patrick. 2022. "The Paradox of Self-Help Expertise: How Unemployed Workers Become Professional Career Coaches." *American Journal of Sociology* 127 (4): 1151–82.

Sherman, Rachel. 2019. *Uneasy Street*. Princeton University Press.

Simmel, Georg. 2006. *Die Großstädte Und Das Geistesleben*. Suhrkamp.

Skeggs, Beverley. 2004. *Class, Self, Culture*. Routledge.

Sklair, Leslie. 2001. *Transnational Capitalist Class*. John Wiley & Sons.

Slaughter, Sheila, and Gary Rhoades. 1997. *Academic Capitalism and the New Economy: Markets, State, and Higher Education*. JHU Press.

Spence, Crawford, and Chris Carter. 2014. "An Exploration of the Professional Habitus in the Big 4 Accounting Firms." *Work, Employment and Society* 28 (6): 946–62.

Spies, Tina. 2017. "Subjektpositionen und positionierungen im diskurs." In *Biographie und diskurs. Theorie und praxis der diskursforschung*, edited by T. Spies and E. Tuider, 69–90. Springer.

Staab, Philipp. 2019. *Digitaler kapitalismus*. Suhrkamp Verlag.

Steinhoff, James. 2021. *Automation and Autonomy: Labour, Capital and Machines in the Artificial Intelligence Industry*. Palgrave Macmillan.

Su, Norman Makoto, Amanda Lazar, and Lilly Irani. 2021. "Critical Affects." *Proceedings of the ACM on Human-Computer Interaction* 5 (CSCW1): 1–27.

Streeck, Wolfgang. 2016. "The Post-Capitalist Interregnum: The Old System Is Dying, but a New Social Order Cannot Yet Be Born." *Juncture* 23 (2): 68–77.

Südkamp, Carolin M., and Sarah E. Dempsey. 2021. "Resistant Transparency and Nonprofit Labor: Challenging Precarity in the Art + Museum Wage Transparency Campaign." *Management Communication Quarterly* 35 (3): 341–67.

Swidler, Ann. 1986. "Culture in Action: Symbols and Strategies." *American Sociological Review* 51 (2): 273–86.

Takhteyev, Yuri. 2012. *Coding Places: Software Practice in a South American City*. MIT Press.

Tan, J. S., and Moira Weigel. 2022. "Organizing in (and Against) a New Cold War: The Case of 996. ICU." In *Digital Work in the Planetary Market*, edited by Mark Graham and Fabian Ferrari, 209–28. MIT Press.

Tan, J. S., Nataliya Nedzhvetskaya, and Emily Mazo. 2023. "Tech Worker Organizing: Understanding the Shift from Occupational to Labor Activism." Cornell University arXiv:2307.15790 [cs.CY], 1–41. https://doi.org/10.48550/arXiv.2307.15790.

Tarnoff, Ben. 2020. "The Making of the Tech Worker Movement." *Logic Magazine*. https://logicmag.io/the-making-of-the-tech-worker-movement/full-text/.

Tarnoff, Ben, and Moira Weigel. 2020. *Voices from the Valley. Tech Workers Talk About What They Do—and How They Do It*. Farraer, Straus and Giroux.

Taylor, Charles. 1992. *Sources of the Self: The Making of the Modern Identity*. Harvard University Press.

Tech Worker Coalition. 2021. "Tech Workers Stand with Gig Workers Against Prop 22 Going National." https://news.techworkerscoalition.org/2021/01/14/issue-2/.

Tech Worker Coalition. 2023. "The Tech Worker Bill of Rights." https://techworkerscoalition.org/bill-of-rights/.

Thaa, Helene. 2020. "Society, Technology and the Future in Tech Development." *Behemoth—A Journal on Civilisation* 13 (1): 57–69.

The Economist. 2023. "Striking Times." *The Economist*, September 16–22. https://www.economist.com/weeklyedition/2023-09-16.

The Economist. 2024. "The War for AI Talent Is Heating Up." *The Economist*, June 15th–21st. https://www.economist.com/weeklyedition/2024-06-15.

Thompson, Edward Palmer. 1966. *The Making of the English Working Class*. Vintage Books.

Timmermans, Stefan, and Iddo Tavory. 2012. "Theory Construction in Qualitative Research: From Grounded Theory to Abductive Analysis." *Sociological Theory* 30 (3): 167–86.

Turner, Fred. 2008. *From Counterculture to Cyberculture: Stewart Brand, the Whole Earth Network and the Rise of Digital Utopianism*. University of Chicago Press.

Ullrich, Christoph. 2022. "Streik an den unikliniken: Landtag schafft möglichkeit zur einigung." June 29. https://www1.wdr.de/nachrichten/landespolitik/streik-uniklinik-landtag-gesetz-100.html.

US BLS. 2023. US Bureau of Labor Statistics. https://www.bls.gov/opub/ted/2023/learning-about-employment-and-wages-in-elementary-and-secondary-schools.htm.

Vaisey, Stephen. 2009. "Motivation and Justification: A Dual-Process Model of Culture in Action." *American Journal of Sociology* 114 (6): 1675–1715.

van Dijck, José, Thomas Poell, and Martijn de Waal. 2018. *The Platform Society: Public Values in a Connective World*. Oxford University Press.

Venturini, Tommaso, and Richard Rogers. 2019. "'API-Based Research' or How Can Digital Sociology and Journalism Studies Learn from the Facebook and Cambridge Analytica Data Breach." *Digital Journalism* 7 (4): 532–40.

Verchere, Alban. 2017. "The Middle-Class Collapse and the Environment." *Ecological Economics* 131:510–23.

Wacquant, Loïc. 2010. "Crafting the Neoliberal State: Workfare, Prisonfare, and Social Insecurity." *Sociological Forum* 25 (2): 197–220.

Waitkus, Nora, and Olaf Groh-Samberg. 2017. "Beyond Meritocracy: Wealth Accumulation in the German Upper Classes." In *New Directions in Elite Studies*, edited by Olav Korsnes, Johan Heilbron, Johs Hjellbrekke, Felix Bühlmann, and Mike Savage, 198–226. Routledge.

Wajcman, Judy. 2006. "New Connections: Social Studies of Science and Technology and Studies of Work." *Work, Employment and Society* 20 (4): 773–86.

Wajcman, Judy. 2018. "How Silicon Valley Sets Time." *New Media & Society* 21 (6): 1272–89.

Waring, Justin, and Asam Latif. 2018. "Of Shepherds, Sheep and Sheepdogs? Governing the Adherent Self Through Complementary and Competing 'Pastorates.'" *Sociology* 52 (5): 1069–86.

Weber, Max. 1904. "Die 'Objektivität' sozialwissenschaftlicher und sozialpolitischer Erkenntnis." *Archiv Für Sozialwissenschaft Und Sozialpolitik* 19 (1): 22–87.

Weber, Max. 1946. "The Social Psychology of the World Religions." In *From Max Weber*, edited by C. Wright Mills and C. C. Gerth, 267–301. Oxford University Press.

Weber, Max. 1949. *The Methodology of The Social Sciences*. Routledge.

Weber, Max. 1968. *Economy and Society*. Bedminster.

Weber, Max. 1994. *Wissenschaft Als Beruf, 1917/1919; Politik Als Beruf, 1919*. Mohr Siebeck.

Weber, Max. 2005. *The Protestant Ethic and the Spirit of Capitalism*. Routledge Classics.

Westheuser, Linus. 2020. "Populism as Symbolic Class Struggle. Homology, Metaphor, and English Ale." *Partecipazione e Conflitto* 13 (1): 256–83.

Westheuser, Linus. 2022. "This Is Not America: Politische Polarisierung in Deutschland Als Schimäre." *Forschungsjournal Soziale Bewegungen* 35 (2): 422–27.

Whyte, William H. 1956. *The Organization Man*. Simon & Schuster.

Williamson, Ben. 2018. "Silicon Startup Schools: Technocracy, Algorithmic Imaginaries and Venture Philanthropy in Corporate Education Reform." *Critical Studies in Education* 59 (2): 218–36.

Wong, Julia Carrie. 2020. "More Than 1,200 Google Workers Condemn Firing of AI Scientist Timnit Gebru." https://www.theguardian.com/technology/2020/dec/04/timnit-gebru-google-ai-fired-diversity-ethics.

Wood, Alex J., Mark Graham, Vili Lehdonvirta, and Isis Hjorth. 2019. "Good Gig, Bad Gig: Autonomy and Algorithmic Control in the Global Gig Economy." *Work, Employment and Society* 33 (1): 56–75.

Wood, Alex J. 2020. *Despotism on Demand: How Power Operates in the Flexible Workplace*. Cornell University Press.

Wright, Erik Olin. 1976. "Class Boundaries in Advanced Capitalist Societies." *New Left Review* I/98 (July–August): 3–41.

Wright, Erik Olin. 2004. *Class Counts: Comparative Studies in Class Analysis*. Cambridge University Press.

Wright, Erik Olin. 2005. *Approaches to Class Analysis*. Cambridge University Press.

Wright, Laurence A., Jonathan Coello, Simon Kemp, and Ian Williams. 2011. "Carbon Footprinting for Climate Change Management in Cities." *Carbon Management* 2 (1): 49–60.

Wu, Tongyu. 2020. "The Labour of Fun: Masculinities and the Organisation of Labour Games in a Modern Workplace." *New Technology, Work and Employment* 35 (3): 336–56.

Wu, Tongyu. 2024. *Play to Submission: Gaming Capitalism in a Tech Firm*. Temple University Press.

Yosifova, Aleksandra. 2023. *How Many Found Jobs After the Tech Layoffs and Where?* 365 Data Science, March 22. https://365datascience.com/trending/the-aftermath-of-the-big-tech-layoffs/.

Yusoff, Kathryn. 2012. "Navigating the Northwest Passage." In *Envisioning Landscapes, Making Worlds*, edited by Stephen Daniels, Dydia DeLyser, Nicholas Entrikin, and Doug Richardson, 331–42. Routledge.

Ziegler, Alexander. 2022. "The Tech Company. On the Neglected Second Nature of Platforms." *Weizenbaum Series* 22: 1–22.

Zuboff, Shoshana. 2019. *The Age of Surveillance Capitalism: The Fight for a Human Future at the New Frontier of Power*. Profile Books.

Index

Publisher contact:
The MIT Press
Massachusetts Institute of Technology
77 Massachusetts Avenue, Cambridge, MA 02139
mitpress.mit.edu

EU Authorised Representative:
Easy Access System Europe, Mustamäe tee 50,
10621 Tallinn, Estonia
gpsr.requests@easproject.com

Printed by Integrated Books International,
United States of America